MCAS Subject Test

Mathematics Grade 4

Student Practice Workbook

+ Two Full-Length MCAS Math Tests

Math Notion

www.MathNotion.com

MCAS Subject Test Mathematics Grade 4

MCAS Subject Test Mathematics Grade 4

Published in the United State of America By

The Math Notion

Web: WWW.MathNotion.com

Email: info@Mathnotion.com

Copyright © 2021 by the Math Notion. All rights reserved. No part of this publication may be reproduced, stored in a retrieval system, or transmitted in any form or by any means, electronic, mechanical, photocopying, recording, scanning, or otherwise, except as permitted under Section 107 or 108 of the 1976 United States Copyright Ac, without permission of the author.

All inquiries should be addressed to the Math Notion.

ISBN: 978-1-63620-094-1

MCAS Subject Test Mathematics Grade 4

The Math Notion

Michael Smith has been a math instructor for over a decade now. He launched the Math Notion. Since 2006, we have devoted our time to both teaching and developing exceptional math learning materials. As a test prep company, we have worked with thousands of students. We have used the feedback of our students to develop a unique study program that can be used by students to drastically improve their math scores fast and effectively. We have more than a thousand Math learning books including:

- **– SAT Math Prep**
- **– ACT Math Prep**
- **– SSAT/ISEE Math Prep**
- **–Mathematics Prep Grade 3 to 8**
- **– Common Core Math Prep**
- **–many Math Education Workbooks, Study Guides, Practice and Exercise Books**

As an experienced Math test preparation company, we have helped many students raise their standardized test scores—and attend the colleges of their dreams: We tutor online and in person, we teach students in large groups, and we provide training materials and textbooks through our website and through Amazon.

You can contact us via email at:

info@Mathnotion.com

MCAS Subject Test Mathematics Grade 4

Get the Targeted Practice You Need to Ace the Massachusetts MCAS Math Test!

MCAS Subject Test Mathematics Grade 4 includes easy-to-follow instructions, helpful examples, and plenty of math practice problems to assist students to master each concept, brush up their problem-solving skills, and create confidence.

The MCAS math practice book provides numerous opportunities to evaluate basic skills along with abundant remediation and intervention activities. It is a skill that permits you to quickly master intricate information and produce better leads in less time.

Students can boost their test-taking skills by taking the book's two practice MCAS Math exams. All test questions answered and explained in detail.

Important Features of the 4th grade MCAS Math Book:

- A **complete review** of MCAS math test topics,
- Over 2,500 practice problems covering all topics tested,
- The most important concepts you need to know,
- Clear and concise, easy-to-follow sections,
- Well designed for enhanced learning and interest,
- Hands-on experience with all question types,
- **2 full-length practice tests** with detailed answer explanations,
- Cost-Effective Pricing,

Powerful math exercises to help you avoid traps and pacing yourself to beat the Massachusetts MCAS test. Students will gain valuable experience and raise their confidence by taking 4th grade math practice tests, learning about test structure, and gaining a deeper understanding of what is tested on the MCAS math grade 4. If ever there was a book to respond to the pressure to increase students' test scores, this is it.

MCAS Subject Test Mathematics Grade 4

WWW.MathNotion.COM

… So Much More Online!

- ✓ FREE Math Lessons
- ✓ More Math Learning Books!
- ✓ Mathematics Worksheets
- ✓ Online Math Tutors

For a PDF Version of This Book

Please Visit WWW.MathNotion.com

MCAS Subject Test Mathematics Grade 4

Contents

Chapter 1 : Place Values and Number Sense 11
   Place Values 12
   Comparing and Ordering Numbers 13
   Numbers in Standard Form 14
   Numbers in Word Form 15
   Roman Numerals 16
   Rounding Numbers 17
   Odd or Even 18
   Answers of Worksheets 19

Chapter 2 : Whole Number Operations 21
   Adding Whole Numbers 22
   Subtracting Whole Numbers 23
   Multiplying Whole Numbers 24
   Dividing Hundreds 25
   Long Division by Two Digits 26
   Division with Remainders 26
   Rounding Whole Numbers 27
   Whole Number Estimation 28
   Answers of Worksheets 29

Chapter 3 : Fractions 31
   Simplifying Fractions 32
   Like Denominators 33
   Compare Fractions with Like Denominators 35
   More Than Two Fractions with Like Denominators 36
   Unlike Denominators 37
   Ordering Fractions 39
   Denominators of 10, 100, and 1000 40
   Multiplying Fractions 42
   Answers of Worksheets 43

Chapter 4 : Mixed Numbers 47
   Fractions to Mixed Numbers 48
   Mixed Numbers to Fractions 49
   Add and Subtract Mixed Numbers 50
   Multiplying Mixed Number 51
   Answers of Worksheets 52

WWW.MathNotion.Com

MCAS Subject Test Mathematics Grade 4

Chapter 5 : Decimals **54**
   Graph Decimals 55
   Adding and Subtracting Decimals 56
   Multiplying and Dividing Decimals 57
   Rounding Decimals 58
   Comparing Decimals 59
   Answers of Worksheets 60

Chapter 6 : Patterns and Algebraic Thinking **62**
   Repeating Patterns 63
   Growing Patterns 64
   Patterns: Numbers 65
   Finding Rules 66
   Algebraic Word Problems 67
   Evaluating Outputs 68
   Answers of Worksheets 69

Chapter 7 : Measurement **71**
   Reference Measurement Units 72
   Metric Length Units 73
   Customary Length Units 73
   Metric Capacity Units 74
   Customary Capacity Units 74
   Metric Weight and Mass Units 75
   Customary Weight and Mass Units 75
   Time 76
   Money Amounts 77
   Money: Word Problems 78
   Answers of Worksheets 79

Chapter 8 : Geometric **81**
   Identifying Angles 82
   Estimate Angle Measurements 83
   Measure Angles with a Protractor 84
   Polygon Names 85
   Classify Triangles 86
   Parallel Sides in Quadrilaterals 87
   Identify Rectangles 88
   Perimeter: Find the Missing Side Lengths 89
   Perimeter and Area of Squares 90
   Perimeter and Area of rectangles 91

MCAS Subject Test Mathematics Grade 4

Find the Area or Missing Side Length of a Rectangle..92
Area and Perimeter: Word Problems..93
Volume of Cubes and Rectangle Prisms 94
Answers of Worksheets 95

Chapter 9 : Data Graphs, and Statistics..97
Bar Graph 98
Tally and Pictographs 99
Dot plots..100
Line Graphs..101
Stem-And-Leaf Plot 102
Coordinate Plane 103
Answers of Worksheets 104

Chapter 10 : Three-Dimensional Figures 106
Identify Three-Dimensional Figures 107
Count Vertices, Edges, and Faces 108
Identify Faces of Three-Dimensional Figures 109
Answers of Worksheets 110

Chapter 11 : Symmetry and Transformations..111
Line Segments 112
Identify Lines of Symmetry 113
Count Lines of Symmetry 114
Parallel, Perpendicular and Intersecting Lines 115
Answers of Worksheets 116

Chapter 12 : MCAS Math Practice Tests..117
MCAS GRADE 4 MAHEMATICS REFRENCE MATERIALS 119
Massachusetts Comprehensive Assessment System Practice Test 1..121
Session 1 122
Session 2 127
Massachusetts Comprehensive Assessment System Practice Test 2..133
Session 1 134
Session 2 138

Chapter 13 : Answers and Explanations..143
Answer Key..143
Practice Test 1..145
Practice Test 2..149

MCAS Subject Test Mathematics Grade 4

Chapter 1 : Place Values and Number Sense

Topics that you'll learn in this chapter:

- ✓ Place Values,
- ✓ Compare and Ordering Numbers,
- ✓ Number in Standard Form
- ✓ Numbers in Word Form,
- ✓ Roman Numerals,
- ✓ Rounding Numbers,
- ✓ Odd or Even,

Place Values

✏ Write numbers in expanded form.

1) Sixty–two ___ + ___

2) fifty–six ___ + ___

3) thirty–one ___ + ___

4) forty–five ___ + ___

5) twenty–eight ___ + ___

✏ Circle the correct choice.

6) The 6 in 56 is in the

   Ones place   tens place   hundreds place

7) The 2 in 25 is in the

   Ones place   tens place   hundreds place

8) The 9 in 918 is in the

   Ones place   tens place   hundreds place

9) The 3 in 537 is in the

   Ones place   tens place   hundreds place

10) The 9 in 289 is in the

   Ones place   tens place   hundreds place

MCAS Subject Test Mathematics Grade 4

Comparing and Ordering Numbers

✏ Use less than, equal to or greater than.

1) 31 _____ 33

2) 57 _____ 49

3) 92 _____ 88

4) 76 _____ 67

5) 43 _____ 43

6) 54 _____ 46

7) 97 _____ 88

8) 42 _____ 36

9) 55 _____ 55

10) 57 _____ 75

11) 28 _____ 38

12) 19 _____ 15

13) 82 _____ 90

14) 78 _____ 84

✏ Order each set numbers from least to greatest.

15) – 18, – 22, 28, – 17, 4   ___, ___, ___, ___, ___, ___

16) 19, –36, 11, – 12, 5   ___, ___, ___, ___, ___, ___

17) 27, – 56, 20, 1, – 27   ___, ___, ___, ___, ___, ___

18) 26, – 96, 2, – 26, 87, –75   ___, ___, ___, ___, ___, ___

19) –10, –71, 70, –26, –59, –39   ___, ___, ___, ___, ___, ___

20) 88, 4, 38, 7, 78, 9   ___, ___, ___, ___, ___, ___

21) 84, 14, 24, 0, 35, 22   ___, ___, ___, ___, ___, ___

MCAS Subject Test Mathematics Grade 4

Numbers in Standard Form

✏ Write the number in standard form.

1) 14 million 154 thousand 8

   _____

2) 89 million 15 thousand 798

   _____

3) 97 million 5 thousand 8

   _____

4) 124 million 2 thousand 2

   _____

5) 50 billion 3 million 5 thousand 4

   _____

6) 34 billion 45 million 578

   _____

7) 94 billion 21 million 51 thousand

   _____

8) 58 billion 708 thousand 120

   _____

9) 59 billion 54 thousand 86

   _____

10) 74 billion 354 thousand 158

    _____

11) 7 billion 13 million 12 thousand 7

    _____

12) 72 billion 450 million 658

    _____

13) 398 million 67 thousand 128

    _____

14) 24 billion 54 million 9 thousand 32

    _____

15) 795 million 458

    _____

16) 38 billion 2 million 54 thousand 9

    _____

MCAS Subject Test Mathematics Grade 4

Numbers in Word Form

✏ Write each number in words.

1) 372 _____

2) 605 _____

3) 550 _____

4) 351 _____

5) 793 _____

6) 647 _____

7) 3,219 _____

8) 5,326 _____

9) 2,842 _____

10) 4,691_____

11) 5,531_____

12) 7,360_____

13) 2,532_____

14) 8,014_____

15) 11,242_____

MCAS Subject Test Mathematics Grade 4

Roman Numerals

✎ Write in Romans numerals.

1	I	11	XI	21	XXI
2	II	12	XII	22	XXII
3	III	13	XIII	23	XXIII
4	IV	14	XIV	24	XXIV
5	V	15	XV	25	XXV
6	VI	16	XVI	26	XXVI
7	VII	17	XVII	27	XXVII
8	VIII	18	XVIII	28	XXVIII
9	IX	19	XIX	29	XXIX
10	X	20	XX	30	XXX

1) 11 _____ 2) 21 _____

3) 24 _____ 4) 16 _____

5) 27 _____ 6) 29 _____

7) 12 _____ 8) 28 _____

9) 15 _____ 10) 20 _____

11) Add 16 + 14 and write in Roman numerals. _____

12) Subtract 34 – 5 and write in Roman numerals. _____

MCAS Subject Test Mathematics Grade 4

Rounding Numbers

※ Round each number to the underlined place value.

1) 3,<u>7</u>93

2) 3,<u>8</u>76

3) 3,4<u>5</u>2

4) 7,1<u>9</u>3

5) 5,2<u>7</u>8

6) 1,4<u>7</u>7

7) 8,<u>3</u>13

8) 24.6<u>8</u>

9) 8<u>4</u>.92

10) 71.<u>3</u>4

11) 66<u>4</u>.7

12) <u>9</u>,135

13) 15.3<u>8</u>1

14) 4,<u>5</u>21

15) 3<u>6</u>.50

16) 4,<u>8</u>19

17) 6,6<u>8</u>5

18) 2,5<u>3</u>8

19) 73.<u>6</u>2

20) 16,<u>5</u>27

21) 2<u>9</u>.720

22) 12,3<u>6</u>6

23) 31,<u>7</u>29

24) 7,8<u>3</u>8

WWW.MathNotion.Com

MCAS Subject Test Mathematics Grade 4

Odd or Even

🖉 Identify whether each number is even or odd.

1) 18 _____   7) 80 _____

2) 27 _____   8) 53 _____

3) 21 _____   9) 58 _____

4) 17 _____   10) 98 _____

5) 67 _____   11) 49 _____

6) 76 _____   12) 113 _____

🖉 Circle the even number in each group.

13) 52, 11, 35, 73, 5, 29   15) 33, 45, 86, 59, 63, 87

14) 13, 15, 113, 87, 71, 18   16) 55, 32, 79, 51, 21, 83

🖉 Circle the odd number in each group.

17) 54, 36, 48, 76, 71, 100   19) 58, 92, 25, 78, 76, 50

18) 32, 56, 40, 74, 98, 67   20) 89, 12, 88, 42, 48, 120

WWW.MathNotion.Com

MCAS Subject Test Mathematics Grade 4

Answers of Worksheets

Place Values

1) 60 + 2
2) 50 + 6
3) 30 + 1
4) 40 + 5
5) 20 + 8
6) ones place
7) tens place
8) hundreds place
9) tens place
10) ones place

Comparing and Ordering Numbers

1) 31 less than 33
2) 57 greater than 49
3) 92 greater than 88
4) 76 greater than 67
5) 43 equals to 43
6) 54 greater than 46
7) 97 greater than 88
8) 42 greater than 36
9) 55 equals to 55
10) 57 less than 75
11) 28 less than 38
12) 19 greater than 15
13) 82 less than 90
14) 78 less than 84
15) –22, –18, –17, 4, 28
16) –36, –12, 5, 11, 19
17) –56, –27, 1, 20, 27
18) –96, –75, –26, 2, 26, 87
19) –71, –59, –39, –26, –10, 70
20) 4, 7, 9, 38, 78, 88
21) 0, 14, 22, 24, 35, 84

Numbers in Standard Form

1) 14,154,008
2) 89,015,798
3) 97,005,008
4) 124,002,002
5) 50,003,005,004
6) 34,054,000,578
7) 94,021,051,000
8) 58,000,708,120
9) 59,000,054,086
10) 74,000,354,158
11) 7,013,012,007
12) 72,450,000,658
13) 398,067,128
14) 24,054,009,032
15) 795,000,458
16) 38,002,054,009

Numbers in Word Form

1) three hundred seventy-two
2) six hundred five
3) five hundred fifty
4) three hundred fifty-one
5) seven hundred ninety-three
6) six hundred forty-seven
7) three thousand, two hundred nineteen
8) five thousand, three hundred twenty-six
9) two thousand, eight hundred forty-two
10) four thousand, six hundred ninety-one
11) five thousand, five hundred thirty-one
12) seven thousand, three hundred sixty
13) two thousand, five hundred thirty-two
14) eight thousand, fourteen

WWW.MathNotion.Com

MCAS Subject Test Mathematics Grade 4

15) eleven thousand, two hundred forty-two

Roman Numerals

1) XI	4) XVI	7) XII	10) XX
2) XXI	5) XXVII	8) XXVIII	11) XXX
3) XXIV	6) XXIX	9) XV	12) XXIX

Rounding Numbers

1) 4,000	7) 8,300	13) 15.380	19) 73.60
2) 4,000	8) 24.70	14) 4,500	20) 16,500
3) 3,450	9) 85.00	15) 37.00	21) 30.00
4) 7,190	10) 71.30	16) 4,800	22) 12,370
5) 5,280	11) 665.00	17) 6,700	23) 31,700
6) 1,480	12) 9,000	18) 2,540	24) 7,840

Odd or Even

1) even	6) even	11) odd	16) 32
2) odd	7) even	12) odd	17) 71
3) odd	8) odd	13) 52	18) 67
4) odd	9) even	14) 18	19) 25
5) odd	10) even	15) 86	20) 89

Chapter 2 : Whole Number Operations

Topics that you'll learn in this chapter:

- ✓ Adding Whole Numbers,
- ✓ Subtracting Whole Numbers,
- ✓ Multiplying Whole Numbers,
- ✓ Dividing Hundreds,
- ✓ Long Division by One Digit,
- ✓ Division with Remainders,
- ✓ Rounding Whole Numbers,
- ✓ Whole Number Estimation,

MCAS Subject Test Mathematics Grade 4

Adding Whole Numbers

✎ Add.

1) 5,763
   + 8,238
   ──────

2) 6,834
   + 4,998
   ──────

3) 3,548
   + 5,693
   ──────

4) 2,769
   +8,872
   ──────

5) 3,196
   +2,936
   ──────

6) 7,009
   + 4,992
   ──────

✎ Find the missing numbers.

7) 3,468 + ___ = 4,102

8) 840 + 2,360 = ___

9) 5,200 + ___ = 7,980

10) 631 + ___ = 2,007

11) ___ + 803 = 3,945

12) ___ + 2,156 = 5,922

13) David sells gems. He finds a diamond in Istanbul and buys it for $4,795. Then, he flies to Cairo and purchases a bigger diamond for the bargain price of $9,633. How much does David spend on the two diamonds? _____

WWW.MathNotion.Com

MCAS Subject Test Mathematics Grade 4

Subtracting Whole Numbers

✎ Subtract.

1) 10,512
   −4,411
   _____

2) 5,204
   − 3,679
   _____

3) 8,520
   − 6,483
   _____

4) 8,001
   − 5,224
   _____

5) 11,916
   − 8,711
   _____

6) 5,005
   −2,008
   _____

✎ Find the missing number.

7) 5,263 − ___ = 2,367

8) 7,198 − ___ = 4,742

9) 8,928 − 3,764 = ___

10) 6,511 − ___ = 3,759

11) 7,003 − 5,489 = ___

12) 8,800 − 5,995 = ___

13) Jackson had $7,189 invested in the stock market until he lost $3,793 on those investments. How much money does he have in the stock market now?

_____

MCAS Subject Test Mathematics Grade 4

Multiplying Whole Numbers

✏️ Find the answers.

1) 2,200
   × 31
   ─────

2) 3,200
   × 22
   ─────

3) 5,790
   × 5
   ─────

4) 5,220
   × 3
   ─────

5) 6,911
   × 3
   ─────

6) 1,998
   × 40
   ─────

7) 2,893
   × 5.5
   ─────

8) 2,254
   × 3.5
   ─────

9) 4,372
   × 4.8
   ─────

10) 3,984
    × 2.75
    ─────

11) 4,900
    × 2.5
    ─────

12) 8,200
    × 4.5
    ─────

MCAS Subject Test Mathematics Grade 4

Dividing Hundreds

✎Find answers.

1) 4,440 ÷ 400

2) 1,600 ÷ 40

3) 9,990 ÷ 90

4) 4,200 ÷ 60

5) 6,400 ÷ 8,000

6) 2,700 ÷ 30

7) 3,333 ÷ 30

8) 558 ÷ 45

9) 2,278 ÷ 85

10) 1,683 ÷ 55

11) 1,582 ÷ 35

12) 9,000 ÷ 600

13) 1,000 ÷ 2,500

14) 44.8 ÷ 20

15) 6,800 ÷ 400

16) 1,500 ÷ 5,000

17) 36.60 ÷ 120

18) 7,700 ÷ 700

19) 5,400 ÷ 600

20) 8,000 ÷ 160

21) 18,000 ÷ 9,000

22) 42,000 ÷ 30

23) 480 ÷ 40

24) 63,000 ÷ 900

MCAS Subject Test Mathematics Grade 4

Long Division by Two Digits

✏ Find the quotient.

1) 18)576

2) 14)952

3) 21)588

4) 23)299

5) 44)748

6) 26)234

7) 16)496

8) 29)1,479

9) 54)1,080

10) 41)1,476

11) 53)2,491

12) 60)2,880

13) 32)2,912

14) 77)8,393

15) 85)3,740

16) 57)4,617

17) 50)9,200

18) 25)15,400

Division with Remainders

✏ Find the quotient with remainder.

1) 14)715

2) 16)2,750

3) 27)4,603

4) 58)2,554

5) 42)7,732

6) 63)6,737

7) 71)9,036

8) 65)8,624

9) 35)5,705

10) 92)13,161

11) 46)12,214

12) 69)42,482

13) 85)6,858

14) 87)34,304

MCAS Subject Test Mathematics Grade 4

Rounding Whole Numbers

Round each number to the underlined place value.

1) 7,<u>5</u>33
2) 9,<u>3</u>74
3) 8,8<u>8</u>3
4) 2,3<u>6</u>8
5) 5,5<u>7</u>7
6) 3,3<u>8</u>1
7) 3,<u>5</u>20
8) 9,3<u>3</u>8
9) 8.<u>5</u>81

10) 33.<u>5</u>7
11) 51.<u>6</u>9
12) 22.<u>1</u>38
13) <u>6</u>,758
14) 11,5<u>5</u>7
15) 8,8<u>3</u>8
16) 5.8<u>8</u>9
17) 1.8<u>6</u>0
18) 25.<u>0</u>70

19) <u>9</u>.332
20) 49.<u>4</u>8
21) 28.<u>8</u>9
22) 24,3<u>7</u>7
23) 52,1<u>5</u>8
24) 13,8<u>8</u>3
25) 9,<u>6</u>09
26) 17,4<u>5</u>1
27) 18,<u>7</u>68

MCAS Subject Test Mathematics Grade 4

Whole Number Estimation

 Estimate the sum by rounding each added to the nearest ten.

1) 875 + 325

2) 985 + 1,452

3) 2,424 + 4,128

4) 1,576 + 6,279

5) 1,247 + 3,863

6) 6,746 + 5,121

7) 3,924 + 6,456

8) 1,785 + 7,164

9) 1,458
   $+$2,442
   ―――

10) 5,689
    $+$4,151
    ―――

11) 8,259
    $+$4,754
    ―――

12) 6,788
    $+$3,954
    ―――

13) 9,123
    $+$4,455
    ―――

14) 6,680
    $+$5,358
    ―――

15) 3,165
    $+$7,124
    ―――

16) 8,859
    $+$6,452
    ―――

WWW.MathNotion.Com

MCAS Subject Test Mathematics Grade 4

Answers of Worksheets

Adding Whole Numbers

1) 14,001
2) 11,832
3) 9,241
4) 11,641
5) 6,132
6) 12,001
7) 634
8) 3,200
9) 2,780
10) 1,376
11) 3,142
12) 3,766
13) $14,428

Subtracting Whole Numbers

1) 6,101
2) 1,525
3) 2,037
4) 2,777
5) 3,205
6) 2,997
7) 2,896
8) 2,456
9) 5,164
10) 2,752
11) 1,514
12) 2,805
13) 3,396

Multiplying Whole Numbers

1) 68,200
2) 70,400
3) 28,950
4) 15,660
5) 20,733
6) 79,920
7) 15,911.5
8) 7,889
9) 20,985.6
10) 10,956
11) 12,250
12) 36,900

Dividing Hundreds

1) 11.1
2) 40
3) 111
4) 70
5) 0.8
6) 90
7) 111.1
8) 12.4
9) 26.8
10) 30.6
11) 45.2
12) 15
13) 0.4
14) 2.24
15) 17
16) 0.3
17) 0.305
18) 11
19) 9
20) 50
21) 2
22) 1,400
23) 12
24) 70

Long Division by Two Digits

1) 32
2) 68
3) 28
4) 13
5) 17
6) 9
7) 31
8) 51
9) 20
10) 36
11) 47
12) 48
13) 91
14) 109
15) 44
16) 81
17) 184
18) 616

MCAS Subject Test Mathematics Grade 4

Division with Remainders

1) 51 R1
2) 171 R14
3) 170 R13
4) 44 R2
5) 184 R4
6) 106 R59
7) 127 R19
8) 132 R44
9) 163 R0
10) 143 R5
11) 265 R24
12) 615 R47
13) 80 R58
14) 394 R26

Rounding Whole Numbers

1) 7,500
2) 9,400
3) 8,880
4) 2,370
5) 5,580
6) 3,380
7) 3,500
8) 9,340
9) 8.60
10) 33.60
11) 51.70
12) 22.100
13) 7,000
14) 11,560
15) 8,840
16) 5.900
17) 1.900
18) 25.100
19) 9.000
20) 49.50
21) 28.90
22) 24,380
23) 52,160
24) 13,880
25) 9,600
26) 17,450
27) 18,800

Whole Number Estimation

1) 1,200
2) 2,440
3) 6,550
4) 7,860
5) 5,110
6) 11,870
7) 10,380
8) 8,950
9) 3,900
10) 9,840
11) 13,010
12) 10,740
13) 13,580
14) 12,040
15) 10,290
16) 15,310

Chapter 3 : Fractions

Topics that you'll learn in this chapter:

- ✓ Simplifying Fractions,
- ✓ Like Denominators,
- ✓ Compare Fractions with Like Denominators,
- ✓ More than two Fractions with Like Denominators,
- ✓ Unlike Denominators,
- ✓ Ordering Fractions,
- ✓ Denominators of 10, 100, and 1000,
- ✓ Multiply Fraction

Simplifying Fractions

Simplify the fractions.

1) $\dfrac{44}{84}$

2) $\dfrac{8}{20}$

3) $\dfrac{12}{16}$

4) $\dfrac{4}{24}$

5) $\dfrac{15}{30}$

6) $\dfrac{9}{63}$

7) $\dfrac{4}{14}$

8) $\dfrac{17}{51}$

9) $\dfrac{24}{30}$

10) $\dfrac{5}{35}$

11) $\dfrac{16}{48}$

12) $\dfrac{33}{22}$

13) $\dfrac{45}{63}$

14) $\dfrac{2.4}{3.2}$

15) $\dfrac{12}{60}$

16) $\dfrac{70}{112}$

17) $\dfrac{2.7}{7.2}$

18) $\dfrac{33}{88}$

19) $\dfrac{1.5}{13.5}$

20) $\dfrac{39}{52}$

21) $\dfrac{5}{45}$

22) $\dfrac{2.1}{4.2}$

MCAS Subject Test Mathematics Grade 4

Like Denominators

✏️ **Add fractions.**

1) $\dfrac{3}{4}+\dfrac{1}{4}$

2) $\dfrac{1}{5}+\dfrac{4}{5}$

3) $\dfrac{4}{9}+\dfrac{7}{9}$

4) $\dfrac{2}{7}+\dfrac{2}{7}$

5) $\dfrac{5}{13}+\dfrac{2}{13}$

6) $\dfrac{1}{14}+\dfrac{4}{14}$

7) $\dfrac{11}{19}+\dfrac{1}{19}$

8) $\dfrac{3}{16}+\dfrac{9}{16}$

9) $\dfrac{3}{10}+\dfrac{1}{10}$

10) $\dfrac{6}{17}+\dfrac{2}{17}$

11) $\dfrac{5}{22}+\dfrac{5}{22}$

12) $\dfrac{7}{35}+\dfrac{11}{35}$

13) $\dfrac{7}{27}+\dfrac{20}{27}$

14) $\dfrac{2}{31}+\dfrac{10}{31}$

15) $\dfrac{5}{23}+\dfrac{3}{23}$

16) $\dfrac{8}{41}+\dfrac{13}{41}$

17) $\dfrac{15}{37}+\dfrac{18}{37}$

18) $\dfrac{2}{51}+\dfrac{7}{51}$

19) $\dfrac{17}{26}+\dfrac{6}{26}$

20) $\dfrac{12}{48}+\dfrac{11}{48}$

21) $\dfrac{11}{29}+\dfrac{8}{29}$

22) $\dfrac{15}{34}+\dfrac{19}{34}$

23) $\dfrac{1}{19}+\dfrac{5}{19}$

24) $\dfrac{3}{53}+\dfrac{4}{53}$

25) $\dfrac{3}{20}+\dfrac{6}{20}$

26) $\dfrac{2}{63}+\dfrac{6}{63}$

27) $\dfrac{6}{38}+\dfrac{1}{38}$

28) $\dfrac{14}{31}+\dfrac{17}{31}$

29) $\dfrac{3}{28}+\dfrac{5}{28}$

30) $\dfrac{2}{37}+\dfrac{15}{37}$

MCAS Subject Test Mathematics Grade 4

✎ Subtract fractions.

1) $\dfrac{8}{9} - \dfrac{4}{9}$

2) $\dfrac{3}{8} - \dfrac{1}{8}$

3) $\dfrac{9}{11} - \dfrac{3}{11}$

4) $\dfrac{9}{14} - \dfrac{4}{14}$

5) $\dfrac{15}{20} - \dfrac{8}{20}$

6) $\dfrac{8}{15} - \dfrac{7}{15}$

7) $\dfrac{11}{19} - \dfrac{9}{19}$

8) $\dfrac{13}{16} - \dfrac{1}{16}$

9) $\dfrac{7}{29} - \dfrac{4}{29}$

10) $\dfrac{14}{23} - \dfrac{7}{23}$

11) $\dfrac{15}{34} - \dfrac{7}{34}$

12) $\dfrac{18}{41} - \dfrac{9}{41}$

13) $\dfrac{17}{39} - \dfrac{16}{39}$

14) $\dfrac{6}{26} - \dfrac{2}{26}$

15) $\dfrac{14}{17} - \dfrac{4}{17}$

16) $\dfrac{33}{55} - \dfrac{20}{55}$

17) $\dfrac{41}{49} - \dfrac{36}{49}$

18) $\dfrac{40}{53} - \dfrac{39}{53}$

19) $\dfrac{27}{37} - \dfrac{17}{37}$

20) $\dfrac{21}{47} - \dfrac{11}{47}$

21) $\dfrac{24}{43} - \dfrac{12}{43}$

22) $\dfrac{13}{19} - \dfrac{12}{19}$

23) $\dfrac{6}{26} - \dfrac{3}{26}$

24) $\dfrac{9}{15} - \dfrac{7}{15}$

25) $\dfrac{8}{39} - \dfrac{3}{39}$

26) $\dfrac{18}{61} - \dfrac{15}{61}$

27) $\dfrac{12}{53} - \dfrac{9}{53}$

28) $\dfrac{75}{76} - \dfrac{74}{76}$

29) $\dfrac{26}{45} - \dfrac{13}{45}$

30) $\dfrac{20}{57} - \dfrac{17}{57}$

MCAS Subject Test Mathematics Grade 4

Compare Fractions with Like Denominators

✎ Evaluate and compare. Write < or > or =.

1) $\dfrac{1}{3} + \dfrac{1}{3}$ — $\dfrac{1}{3}$

2) $\dfrac{3}{6} + \dfrac{3}{6}$ — $\dfrac{5}{6}$

3) $\dfrac{8}{9} - \dfrac{4}{9}$ — $\dfrac{7}{9}$

4) $\dfrac{4}{11} + \dfrac{5}{11}$ — $\dfrac{7}{11}$

5) $\dfrac{9}{14} - \dfrac{8}{14}$ — $\dfrac{5}{14}$

6) $\dfrac{11}{17} - \dfrac{3}{17}$ — $\dfrac{6}{17}$

7) $\dfrac{11}{21} + \dfrac{2}{21}$ — $\dfrac{10}{21}$

8) $\dfrac{8}{32} + \dfrac{6}{32}$ — $\dfrac{9}{32}$

9) $\dfrac{25}{29} - \dfrac{16}{29}$ — $\dfrac{11}{29}$

10) $\dfrac{28}{41} + \dfrac{13}{41}$ — $\dfrac{27}{41}$

11) $\dfrac{18}{35} - \dfrac{11}{35}$ — $\dfrac{22}{35}$

12) $\dfrac{32}{47} - \dfrac{22}{47}$ — $\dfrac{11}{47}$

13) $\dfrac{14}{27} + \dfrac{13}{27}$ — $\dfrac{24}{27}$

14) $\dfrac{34}{52} - \dfrac{11}{52}$ — $\dfrac{21}{52}$

15) $\dfrac{43}{56} - \dfrac{24}{56}$ — $\dfrac{27}{56}$

16) $\dfrac{27}{71} + \dfrac{25}{71}$ — $\dfrac{48}{71}$

WWW.MathNotion.Com

MCAS Subject Test Mathematics Grade 4

More Than Two Fractions with Like Denominators

🖎 Add fractions.

1) $\dfrac{5}{9} + \dfrac{2}{9} + \dfrac{2}{9}$

2) $\dfrac{4}{6} + \dfrac{1}{6} + \dfrac{1}{6}$

3) $\dfrac{2}{17} + \dfrac{4}{17} + \dfrac{2}{17}$

4) $\dfrac{1}{5} + \dfrac{1}{5} + \dfrac{1}{5}$

5) $\dfrac{7}{18} + \dfrac{2}{18} + \dfrac{3}{18}$

6) $\dfrac{3}{27} + \dfrac{5}{27} + \dfrac{2}{27}$

7) $\dfrac{4}{33} + \dfrac{4}{33} + \dfrac{4}{33}$

8) $\dfrac{8}{23} + \dfrac{6}{23} + \dfrac{2}{23}$

9) $\dfrac{13}{41} + \dfrac{2}{41} + \dfrac{8}{41}$

10) $\dfrac{6}{35} + \dfrac{9}{35} + \dfrac{20}{35}$

11) $\dfrac{1}{37} + \dfrac{5}{37} + \dfrac{5}{37}$

12) $\dfrac{4}{43} + \dfrac{9}{43} + \dfrac{8}{43}$

13) $\dfrac{4}{51} + \dfrac{10}{51} + \dfrac{7}{51}$

14) $\dfrac{5}{26} + \dfrac{13}{26} + \dfrac{6}{26}$

15) $\dfrac{5}{64} + \dfrac{4}{64} + \dfrac{2}{64}$

16) $\dfrac{1}{73} + \dfrac{5}{73} + \dfrac{6}{73}$

WWW.MathNotion.Com

MCAS Subject Test Mathematics Grade 4

Unlike Denominators

Add fraction.

1) $\dfrac{2}{9} + \dfrac{3}{4}$

2) $\dfrac{1}{4} + \dfrac{3}{5}$

3) $\dfrac{1}{16} + \dfrac{3}{4}$

4) $\dfrac{3}{8} + \dfrac{1}{7}$

5) $\dfrac{1}{3} + \dfrac{2}{4}$

6) $\dfrac{1}{6} + \dfrac{3}{7}$

7) $\dfrac{5}{18} + \dfrac{4}{6}$

8) $\dfrac{1}{12} + \dfrac{5}{6}$

9) $\dfrac{5}{27} + \dfrac{1}{9}$

10) $\dfrac{1}{6} + \dfrac{7}{24}$

11) $\dfrac{3}{5} + \dfrac{1}{8}$

12) $\dfrac{11}{42} + \dfrac{3}{7}$

13) $\dfrac{7}{20} + \dfrac{1}{3}$

14) $\dfrac{1}{45} + \dfrac{3}{5}$

15) $\dfrac{3}{32} + \dfrac{5}{8}$

16) $\dfrac{3}{48} + \dfrac{5}{6}$

17) $\dfrac{5}{12} + \dfrac{1}{6}$

18) $\dfrac{1}{34} + \dfrac{3}{17}$

19) $\dfrac{4}{9} + \dfrac{7}{54}$

20) $\dfrac{13}{56} + \dfrac{4}{7}$

21) $\dfrac{3}{12} + \dfrac{2}{3}$

22) $\dfrac{4}{33} + \dfrac{5}{11}$

WWW.MathNotion.Com

MCAS Subject Test Mathematics Grade 4

✏️ Subtract fractions.

1) $\dfrac{8}{9} - \dfrac{1}{2}$

2) $\dfrac{2}{3} - \dfrac{3}{10}$

3) $\dfrac{1}{6} - \dfrac{1}{9}$

4) $\dfrac{7}{8} - \dfrac{1}{4}$

5) $\dfrac{3}{4} - \dfrac{1}{28}$

6) $\dfrac{11}{30} - \dfrac{3}{15}$

7) $\dfrac{11}{18} - \dfrac{5}{9}$

8) $\dfrac{5}{13} - \dfrac{3}{26}$

9) $\dfrac{17}{35} - \dfrac{2}{7}$

10) $\dfrac{5}{6} - \dfrac{12}{36}$

11) $\dfrac{5}{9} - \dfrac{1}{27}$

12) $\dfrac{3}{5} - \dfrac{1}{8}$

13) $\dfrac{2}{3} - \dfrac{3}{5}$

14) $\dfrac{7}{8} - \dfrac{3}{7}$

15) $\dfrac{5}{9} - \dfrac{13}{45}$

16) $\dfrac{3}{4} - \dfrac{5}{36}$

17) $\dfrac{39}{49} - \dfrac{5}{7}$

18) $\dfrac{3}{11} - \dfrac{3}{22}$

19) $\dfrac{17}{48} - \dfrac{4}{12}$

20) $\dfrac{2}{3} - \dfrac{4}{13}$

21) $\dfrac{5}{8} - \dfrac{19}{72}$

22) $\dfrac{3}{5} - \dfrac{1}{12}$

WWW.MathNotion.Com

MCAS Subject Test Mathematics Grade 4

Ordering Fractions

✏ Order the fractions from least to greatest.

1) $\frac{1}{5}, \frac{1}{11}, \frac{1}{8}, \frac{1}{3}$ ____, ____, ____, ____

2) $\frac{1}{9}, \frac{1}{18}, \frac{2}{4}, \frac{1}{5}$ ____, ____, ____, ____

3) $\frac{4}{7}, \frac{1}{7}, \frac{6}{21}, \frac{15}{21}$ ____, ____, ____, ____

4) $\frac{1}{2}, \frac{1}{3}, \frac{4}{9}, \frac{5}{18}$ ____, ____, ____, ____

5) $\frac{4}{9}, \frac{3}{4}, \frac{7}{36}, \frac{1}{6}$ ____, ____, ____, ____

✏ Order the fractions from greatest to least.

6) $\frac{3}{4}, \frac{4}{7}, \frac{3}{10}, \frac{5}{13}$ ____, ____, ____, ____

7) $\frac{5}{11}, \frac{5}{6}, \frac{2}{5}, \frac{1}{3}$ ____, ____, ____, ____

8) $\frac{7}{8}, \frac{1}{6}, \frac{3}{4}, \frac{5}{15}$ ____, ____, ____, ____

9) $\frac{4}{7}, \frac{2}{3}, \frac{11}{25}, \frac{13}{33}$ ____, ____, ____, ____

10) $\frac{18}{20}, \frac{15}{16}, \frac{14}{18}, \frac{5}{12}$ ____, ____, ____, ____

MCAS Subject Test Mathematics Grade 4

Denominators of 10, 100, and 1000

Add fractions.

1) $\dfrac{7}{10} + \dfrac{13}{100}$

2) $\dfrac{1}{10} + \dfrac{10}{100}$

3) $\dfrac{15}{100} + \dfrac{1}{1,000}$

4) $\dfrac{56}{100} + \dfrac{3}{10}$

5) $\dfrac{50}{1,000} + \dfrac{7}{10}$

6) $\dfrac{6}{10} + \dfrac{30}{1,000}$

7) $\dfrac{9}{100} + \dfrac{3}{10}$

8) $\dfrac{5}{10} + \dfrac{50}{100}$

9) $\dfrac{48}{100} + \dfrac{6}{10}$

10) $\dfrac{70}{100} + \dfrac{2}{10}$

11) $\dfrac{80}{100} + \dfrac{200}{1,000}$

12) $\dfrac{30}{100} + \dfrac{4}{10}$

13) $\dfrac{9}{100} + \dfrac{7}{10}$

14) $\dfrac{25}{100} + \dfrac{6}{10}$

15) $\dfrac{15}{100} + \dfrac{8}{10}$

16) $\dfrac{3}{10} + \dfrac{31}{100}$

17) $\dfrac{8}{10} + \dfrac{11}{100}$

18) $\dfrac{34}{100} + \dfrac{6}{10}$

WWW.MathNotion.Com

MCAS Subject Test Mathematics Grade 4

🖉 Subtract fractions.

1) $\dfrac{8}{10} - \dfrac{20}{100}$

2) $\dfrac{5}{10} - \dfrac{47}{100}$

3) $\dfrac{12}{100} - \dfrac{60}{1,000}$

4) $\dfrac{6}{10} - \dfrac{50}{100}$

5) $\dfrac{3}{10} - \dfrac{23}{100}$

6) $\dfrac{70}{100} - \dfrac{250}{1,000}$

7) $\dfrac{4}{10} - \dfrac{350}{1,000}$

8) $\dfrac{70}{100} - \dfrac{3}{10}$

9) $\dfrac{40}{100} - \dfrac{3}{10}$

10) $\dfrac{6}{10} - \dfrac{180}{1,000}$

11) $\dfrac{93}{100} - \dfrac{5}{10}$

12) $\dfrac{65}{100} - \dfrac{4}{10}$

13) $\dfrac{80}{100} - \dfrac{6}{10}$

14) $\dfrac{90}{100} - \dfrac{5}{10}$

15) $\dfrac{200}{1,000} - \dfrac{1}{10}$

16) $\dfrac{90}{100} - \dfrac{7}{10}$

17) $\dfrac{900}{1,000} - \dfrac{40}{100}$

18) $\dfrac{60}{100} - \dfrac{3}{10}$

MCAS Subject Test Mathematics Grade 4

Multiplying Fractions

Find the product.

1) $\dfrac{4}{5} \times \dfrac{2}{6} =$

2) $\dfrac{4}{22} \times \dfrac{5}{8} =$

3) $\dfrac{8}{30} \times \dfrac{12}{16} =$

4) $\dfrac{9}{14} \times \dfrac{21}{36} =$

5) $\dfrac{14}{15} \times \dfrac{5}{7} =$

6) $\dfrac{16}{19} \times \dfrac{3}{4} =$

7) $\dfrac{4}{9} \times \dfrac{9}{8} =$

8) $\dfrac{47}{85} \times 0 =$

9) $\dfrac{5}{8} \times \dfrac{16}{6} =$

10) $\dfrac{28}{15} \times \dfrac{5}{7} =$

11) $\dfrac{32}{24} \times \dfrac{12}{16} =$

12) $\dfrac{6}{42} \times \dfrac{7}{36} =$

13) $\dfrac{13}{8} \times \dfrac{12}{4} =$

14) $\dfrac{10}{9} \times \dfrac{6}{5} =$

15) $\dfrac{35}{56} \times \dfrac{8}{7} =$

16) $\dfrac{16}{18} \times 9 =$

17) $\dfrac{5}{22} \times \dfrac{44}{15} =$

18) $\dfrac{10}{18} \times \dfrac{9}{20} =$

19) $\dfrac{7}{11} \times \dfrac{8}{21} =$

20) $\dfrac{26}{24} \times \dfrac{8}{52} =$

21) $\dfrac{6}{17} \times \dfrac{1}{12} =$

22) $\dfrac{20}{9} \times \dfrac{6}{100} =$

23) $\dfrac{8}{14} \times \dfrac{7}{72} =$

24) $\dfrac{50}{100} \times \dfrac{300}{400} =$

WWW.MathNotion.Com

MCAS Subject Test Mathematics Grade 4

Answers of Worksheets

Simplifying Fractions

1) $\frac{11}{21}$
2) $\frac{2}{5}$
3) $\frac{3}{4}$
4) $\frac{1}{6}$
5) $\frac{1}{2}$
6) $\frac{1}{7}$
7) $\frac{2}{7}$
8) $\frac{1}{3}$
9) $\frac{4}{5}$
10) $\frac{1}{7}$
11) $\frac{1}{3}$
12) $\frac{3}{2}$
13) $\frac{5}{7}$
14) $\frac{3}{4}$
15) $\frac{1}{5}$
16) $\frac{5}{8}$
17) $\frac{3}{8}$
18) $\frac{3}{8}$
19) $\frac{1}{9}$
20) $\frac{3}{4}$
21) $\frac{1}{9}$
22) $\frac{1}{2}$

Like Denominators (addition)

1) 1
2) 1
3) $\frac{11}{9}$
4) $\frac{4}{7}$
5) $\frac{7}{13}$
6) $\frac{5}{14}$
7) $\frac{12}{19}$
8) $\frac{3}{4}$
9) $\frac{2}{5}$
10) $\frac{8}{17}$
11) $\frac{5}{11}$
12) $\frac{18}{35}$
13) 1
14) $\frac{12}{31}$
15) $\frac{8}{23}$
16) $\frac{21}{41}$
17) $\frac{33}{37}$
18) $\frac{3}{17}$
19) $\frac{23}{26}$
20) $\frac{23}{48}$
21) $\frac{19}{29}$
22) 1
23) $\frac{6}{19}$
24) $\frac{7}{53}$
25) $\frac{9}{20}$
26) $\frac{8}{63}$
27) $\frac{7}{38}$
28) 1
29) $\frac{2}{7}$
30) $\frac{17}{37}$

Like Denominators (Subtraction)

1) $\frac{4}{9}$
2) $\frac{1}{4}$
3) $\frac{6}{11}$
4) $\frac{5}{14}$
5) $\frac{7}{20}$
6) $\frac{1}{15}$
7) $\frac{2}{19}$
8) $\frac{3}{4}$
9) $\frac{3}{29}$
10) $\frac{7}{23}$
11) $\frac{8}{34}$
12) $\frac{9}{41}$
13) $\frac{1}{39}$
14) $\frac{2}{13}$
15) $\frac{10}{17}$
16) $\frac{13}{55}$
17) $\frac{5}{49}$
18) $\frac{1}{53}$
19) $\frac{10}{37}$
20) $\frac{10}{47}$
21) $\frac{12}{43}$
22) $\frac{1}{19}$
23) $\frac{3}{26}$
24) $\frac{2}{15}$
25) $\frac{5}{39}$
26) $\frac{3}{61}$
27) $\frac{3}{53}$
28) $\frac{1}{76}$

WWW.MathNotion.Com

MCAS Subject Test Mathematics Grade 4

29) $\frac{13}{45}$ 30) $\frac{1}{19}$

Compare Fractions with Like Denominators

1) $\frac{2}{3} > \frac{1}{3}$ 5) $\frac{1}{14} < \frac{5}{14}$ 9) $\frac{9}{29} < \frac{11}{29}$ 13) $1 > \frac{24}{27}$

2) $1 > \frac{5}{6}$ 6) $\frac{8}{17} > \frac{6}{17}$ 10) $1 > \frac{27}{41}$ 14) $\frac{23}{52} > \frac{21}{52}$

3) $\frac{4}{9} < \frac{7}{9}$ 7) $\frac{13}{21} > \frac{10}{21}$ 11) $\frac{7}{35} < \frac{22}{35}$ 15) $\frac{19}{56} < \frac{27}{56}$

4) $\frac{9}{11} > \frac{7}{11}$ 8) $\frac{14}{32} > \frac{9}{32}$ 12) $\frac{10}{47} < \frac{11}{47}$ 16) $\frac{52}{71} > \frac{48}{71}$

More Than Two Fractions with Like Denominators

1) 1   5) $\frac{2}{3}$ 9) $\frac{23}{41}$ 13) $\frac{7}{17}$

2) 1   6) $\frac{10}{27}$ 10) 1   14) $\frac{12}{13}$

3) $\frac{8}{17}$ 7) $\frac{4}{11}$ 11) $\frac{11}{37}$ 15) $\frac{11}{64}$

4) $\frac{3}{5}$ 8) $\frac{16}{23}$ 12) $\frac{21}{43}$ 16) $\frac{12}{73}$

Unlike Denominators (Addition)

1) $\frac{35}{36}$ 7) $\frac{17}{18}$ 13) $\frac{41}{60}$ 19) $\frac{31}{54}$

2) $\frac{17}{20}$ 8) $\frac{11}{12}$ 14) $\frac{28}{45}$ 20) $\frac{45}{56}$

3) $\frac{13}{16}$ 9) $\frac{8}{27}$ 15) $\frac{23}{32}$ 21) $\frac{11}{12}$

4) $\frac{29}{56}$ 10) $\frac{11}{24}$ 16) $\frac{43}{48}$ 22) $\frac{19}{33}$

5) $\frac{5}{6}$ 11) $\frac{29}{40}$ 17) $\frac{7}{12}$

6) $\frac{25}{42}$ 12) $\frac{29}{42}$ 18) $\frac{7}{34}$

Unlike Denominators (Subtraction)

1) $\frac{7}{18}$ 7) $\frac{1}{18}$ 13) $\frac{1}{15}$ 19) $\frac{1}{48}$

2) $\frac{11}{30}$ 8) $\frac{7}{26}$ 14) $\frac{25}{56}$ 20) $\frac{14}{39}$

3) $\frac{1}{18}$ 9) $\frac{1}{5}$ 15) $\frac{4}{15}$ 21) $\frac{13}{36}$

4) $\frac{5}{8}$ 10) $\frac{1}{2}$ 16) $\frac{11}{18}$ 22) $\frac{31}{60}$

5) $\frac{5}{7}$ 11) $\frac{14}{27}$ 17) $\frac{4}{49}$

6) $\frac{1}{6}$ 12) $\frac{19}{40}$ 18) $\frac{3}{22}$

MCAS Subject Test Mathematics Grade 4

Ordering Fractions

1) $\frac{1}{11}, \frac{1}{8}, \frac{1}{5}, \frac{1}{3}$

2) $\frac{1}{18}, \frac{1}{9}, \frac{1}{5}, \frac{2}{4}$

3) $\frac{1}{7}, \frac{6}{21}, \frac{4}{7}, \frac{15}{21}$

4) $\frac{5}{18}, \frac{1}{3}, \frac{4}{9}, \frac{1}{2}$

5) $\frac{1}{6}, \frac{7}{36}, \frac{4}{9}, \frac{3}{4}$

6) $\frac{3}{4}, \frac{4}{7}, \frac{5}{13}, \frac{3}{10}$

7) $\frac{5}{6}, \frac{5}{11}, \frac{2}{5}, \frac{1}{3}$

8) $\frac{7}{8}, \frac{3}{4}, \frac{5}{15}, \frac{1}{6}$

9) $\frac{2}{3}, \frac{4}{7}, \frac{11}{25}, \frac{13}{33}$

10) $\frac{15}{16}, \frac{18}{20}, \frac{14}{18}, \frac{5}{12}$

Denominators of 10, 100, and 1000

1) $\frac{83}{100}$

2) $\frac{1}{5}$

3) $\frac{151}{1,000}$

4) $\frac{43}{50}$

5) $\frac{3}{4}$

6) $\frac{63}{100}$

7) $\frac{39}{100}$

8) 1

9) $\frac{27}{25}$

10) $\frac{9}{10}$

11) 1

12) $\frac{7}{10}$

13) $\frac{79}{100}$

14) $\frac{17}{20}$

15) $\frac{19}{20}$

16) $\frac{61}{100}$

17) $\frac{91}{100}$

18) $\frac{47}{50}$

Denominators of 10, 100, and 1000 (Subtract)

1) $\frac{3}{5}$

2) $\frac{3}{100}$

3) $\frac{3}{50}$

4) $\frac{1}{10}$

5) $\frac{7}{100}$

6) $\frac{9}{20}$

7) $\frac{1}{20}$

8) $\frac{2}{5}$

9) $\frac{1}{10}$

10) $\frac{21}{50}$

11) $\frac{43}{100}$

12) $\frac{1}{4}$

13) $\frac{1}{5}$

14) $\frac{2}{5}$

15) $\frac{1}{10}$

16) $\frac{1}{5}$

17) $\frac{1}{2}$

18) $\frac{3}{10}$

Multiplying Fractions

1) $\frac{4}{15}$

2) $\frac{5}{44}$

3) $\frac{1}{5}$

4) $\frac{3}{8}$

5) $\frac{2}{3}$

6) $\frac{12}{19}$

7) $\frac{1}{2}$

8) 0

9) $\frac{5}{3}$

10) $\frac{4}{3}$

11) 1

12) $\frac{1}{36}$

13) $\frac{39}{8}$

14) $\frac{4}{3}$

15) $\frac{5}{7}$

16) 8

17) $\frac{2}{3}$

18) $\frac{1}{4}$

19) $\frac{8}{33}$

20) $\frac{1}{6}$

21) $\frac{1}{34}$

22) $\frac{2}{15}$

23) $\frac{1}{18}$

24) $\frac{3}{8}$

WWW.MathNotion.Com

MCAS Subject Test Mathematics Grade 4

Chapter 4:
Mixed Numbers

Topics that you'll learn in this chapter:

- ✓ Fractions to Mixed Numbers,
- ✓ Mixed Numbers to Fractions,
- ✓ Add and Subtract Mixed Numbers,
- ✓ Multiply Mixed Numbers

MCAS Subject Test Mathematics Grade 4

Fractions to Mixed Numbers

✎ Convert fractions to mixed numbers.

1) $\dfrac{9}{5}$ 11) $\dfrac{41}{9}$

2) $\dfrac{11}{3}$ 12) $\dfrac{45}{12}$

3) $\dfrac{39}{8}$ 13) $\dfrac{17}{5}$

4) $\dfrac{27}{11}$ 14) $\dfrac{29}{6}$

5) $\dfrac{7}{2}$ 15) $\dfrac{13}{4}$

6) $\dfrac{43}{4}$ 16) $\dfrac{15}{7}$

7) $\dfrac{49}{9}$ 17) $\dfrac{65}{7}$

8) $\dfrac{15}{4}$ 18) $\dfrac{59}{8}$

9) $\dfrac{37}{7}$ 19) $\dfrac{25}{4}$

10) $\dfrac{19}{7}$ 20) $\dfrac{17}{8}$

MCAS Subject Test Mathematics Grade 4

Mixed Numbers to Fractions

🖎 Convert to fraction.

1) $3\frac{3}{5}$

2) $1\frac{1}{3}$

3) $4\frac{2}{5}$

4) $4\frac{2}{8}$

5) $2\frac{1}{5}$

6) $2\frac{8}{11}$

7) $4\frac{4}{7}$

8) $3\frac{7}{12}$

9) $2\frac{1}{3}$

10) $7\frac{5}{7}$

11) $2\frac{7}{10}$

12) $3\frac{4}{9}$

13) $1\frac{5}{8}$

14) $4\frac{3}{11}$

15) $3\frac{4}{7}$

16) $5\frac{2}{8}$

17) $7\frac{1}{7}$

18) $13\frac{1}{2}$

19) $4\frac{2}{7}$

20) $5\frac{2}{10}$

21) $12\frac{1}{3}$

22) $7\frac{1}{8}$

MCAS Subject Test Mathematics Grade 4

Add and Subtract Mixed Numbers

✍ Add mixed numbers.

1) $3\frac{2}{5} + 8\frac{1}{5}$

2) $3\frac{2}{3} + 4\frac{1}{2}$

3) $6\frac{2}{7} + 2\frac{3}{7}$

4) $4\frac{2}{5} + 3\frac{1}{4}$

5) $8\frac{3}{4} - 2\frac{1}{2}$

6) $6\frac{5}{12} - 4\frac{1}{4}$

7) $5\frac{3}{8} - 3\frac{7}{8}$

8) $6\frac{1}{4} - 2\frac{15}{16}$

9) $9\frac{23}{28} - 4\frac{17}{28}$

10) $6\frac{1}{6} + 6\frac{2}{3}$

11) $4\frac{2}{9} + 5\frac{5}{9}$

12) $2\frac{1}{4} + 7\frac{4}{7}$

13) $7\frac{1}{5} - 3\frac{3}{5}$

14) $3\frac{1}{6} + 2\frac{3}{7}$

15) $2\frac{1}{3} + 4\frac{1}{4}$

16) $4\frac{1}{4} - 1\frac{2}{5}$

17) $\frac{1}{3} + 6\frac{1}{6}$

18) $2\frac{3}{5} + 2\frac{1}{10}$

WWW.MathNotion.Com

MCAS Subject Test Mathematics Grade 4

Multiplying Mixed Number

Multiply. Reduce to lowest terms.

1) $2\frac{3}{5} \times 1\frac{3}{4} =$

2) $1\frac{5}{6} \times 1\frac{1}{3} =$

3) $2\frac{3}{5} \times 1\frac{1}{7} =$

4) $3\frac{1}{7} \times 2\frac{1}{2} =$

5) $4\frac{3}{4} \times 1\frac{1}{4} =$

6) $3\frac{1}{2} \times 1\frac{4}{5} =$

7) $3\frac{3}{4} \times 1\frac{1}{2} =$

8) $5\frac{2}{3} \times 3\frac{1}{3} =$

9) $3\frac{2}{3} \times 3\frac{1}{2} =$

10) $2\frac{1}{3} \times 3\frac{1}{2} =$

11) $4\frac{3}{4} \times 3\frac{2}{3} =$

12) $3\frac{2}{4} \times 3\frac{1}{6} =$

13) $2\frac{2}{5} \times 1\frac{1}{3} =$

14) $2\frac{1}{3} \times 1\frac{1}{6} =$

15) $2\frac{2}{3} \times 3\frac{1}{2} =$

16) $2\frac{1}{8} \times 2\frac{2}{5} =$

17) $2\frac{1}{4} \times 1\frac{2}{3} =$

18) $2\frac{3}{5} \times 1\frac{1}{4} =$

19) $2\frac{3}{5} \times 1\frac{5}{6} =$

20) $3\frac{3}{5} \times 2\frac{3}{4} =$

21) $3\frac{3}{4} \times 1\frac{1}{3} =$

22) $2\frac{5}{8} \times 3\frac{1}{4} =$

WWW.MathNotion.Com

MCAS Subject Test Mathematics Grade 4

Answers of Worksheets

Fractions to Mixed Numbers

1) $1\frac{4}{5}$ 6) $10\frac{3}{4}$ 11) $4\frac{5}{9}$ 16) $2\frac{1}{7}$

2) $3\frac{2}{3}$ 7) $5\frac{4}{9}$ 12) $3\frac{9}{12}$ 17) $9\frac{2}{7}$

3) $4\frac{7}{8}$ 8) $3\frac{3}{4}$ 13) $3\frac{2}{5}$ 18) $7\frac{3}{8}$

4) $2\frac{5}{11}$ 9) $5\frac{2}{7}$ 14) $4\frac{5}{6}$ 19) $6\frac{1}{4}$

5) $3\frac{1}{2}$ 10) $2\frac{5}{7}$ 15) $3\frac{1}{4}$ 20) $2\frac{1}{8}$

Mixed Numbers to Fractions

1) $\frac{18}{5}$ 7) $\frac{32}{7}$ 13) $\frac{13}{8}$ 19) $\frac{30}{7}$

2) $\frac{4}{3}$ 8) $\frac{43}{12}$ 14) $\frac{47}{11}$ 20) $\frac{52}{10}$

3) $\frac{22}{5}$ 9) $\frac{7}{3}$ 15) $\frac{25}{7}$ 21) $\frac{37}{3}$

4) $\frac{34}{8}$ 10) $\frac{54}{7}$ 16) $\frac{42}{8}$ 22) $\frac{57}{8}$

5) $\frac{11}{5}$ 11) $\frac{27}{10}$ 17) $\frac{50}{7}$

6) $\frac{30}{11}$ 12) $\frac{31}{9}$ 18) $\frac{27}{2}$

Add and Subtract Mixed Numbers

1) $11\frac{3}{5}$ 6) $2\frac{1}{6}$ 11) $9\frac{7}{9}$ 16) $2\frac{17}{20}$

2) $8\frac{1}{6}$ 7) $1\frac{1}{2}$ 12) $9\frac{23}{28}$ 17) $6\frac{1}{2}$

3) $8\frac{5}{7}$ 8) $3\frac{5}{16}$ 13) $3\frac{3}{5}$ 18) $4\frac{7}{10}$

4) $7\frac{13}{20}$ 9) $5\frac{3}{14}$ 14) $5\frac{25}{42}$

5) $6\frac{1}{4}$ 10) $12\frac{5}{6}$ 15) $6\frac{7}{12}$

Multiplying Mixed Number

1) $4\frac{11}{20}$ 4) $7\frac{6}{7}$ 7) $5\frac{5}{8}$

2) $2\frac{4}{9}$ 5) $5\frac{15}{16}$ 8) $18\frac{8}{9}$

3) $2\frac{34}{35}$ 6) $6\frac{3}{10}$ 9) $12\frac{5}{6}$

WWW.MathNotion.Com

MCAS Subject Test Mathematics Grade 4

10) $8\frac{1}{6}$

11) $17\frac{5}{12}$

12) $11\frac{1}{12}$

13) $3\frac{1}{5}$

14) $2\frac{13}{18}$

15) $9\frac{1}{3}$

16) $5\frac{1}{10}$

17) $3\frac{3}{4}$

18) $3\frac{1}{4}$

19) $4\frac{23}{30}$

20) $9\frac{9}{10}$

21) 5

22) $8\frac{17}{32}$

MCAS Subject Test Mathematics Grade 4

Chapter 5 : Decimals

Topics that you'll learn in this chapter:

- ✓ Graph Decimal
- ✓ Adding and Subtracting Decimals,
- ✓ Multiplying and Dividing Decimals,
- ✓ Round decimals,
- ✓ Comparing Decimals,

MCAS Subject Test Mathematics Grade 4

Graph Decimals

✎ Write the decimals indicated by the arrows.

1)

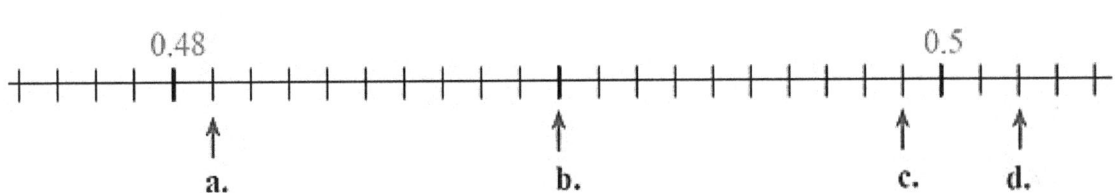

a. _____ b. _____ c. _____ d. _____

2)

a. _____ b. _____ c. _____ d. _____

3)

a. _____ b. _____ c. _____ d. _____

4)

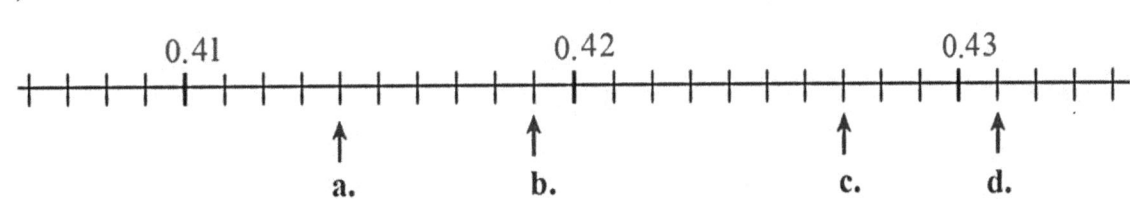

a. _____ b. _____ c. _____ d. _____

WWW.MathNotion.Com

MCAS Subject Test Mathematics Grade 4

Adding and Subtracting Decimals

✎ Add and subtract decimals.

1) 24.19
   − 15.42
   ─────

2) 42.23
   + 25.42
   ─────

3) 72.54
   + 11.28
   ─────

4) 57.45
   − 24.75
   ─────

5) 43.57
   + 54.85
   ─────

6) 86.68
   − 54.12
   ─────

✎ Solve.

7) ____ + 2.7 = 8.1

8) 6.4 + ____ = 12.8

9) 7.9 + ____ = 17

10) 4.6 + ____ = 15.3

11) ____ + 9.4 = 15

12) ____ + 8.24 = 13.54

✎ Order each set of numbers from least to greatest.

1) 0.4, 0.67, 0.44, 0.73, 0.51   ___, ___, ___, ___, ___, ___

2) 3.9, 6.1, 4.28, 7.02, 4.65   ___, ___, ___, ___, ___, ___

3) 1.9, 1.04, 0.79, 0.72, 0.09   ___, ___, ___, ___, ___, ___

4) 2.6, 5.2, 1.9, 4.01, 1.99, 3.2   ___, ___, ___, ___, ___, ___

5) 4.2, 6.1, 3.8, 5.7, 2.1, 2.8   ___, ___, ___, ___, ___, ___

6) 0.56, 0.87, 0.14, 1.24, 3.1   ___, ___, ___, ___, ___, ___

MCAS Subject Test Mathematics Grade 4

Multiplying and Dividing Decimals

✏️ Find each product.

1) 1.5
   × 2.1
   -----

2) 4.6
   × 3.4
   -----

3) 6.3
   × 2.5
   -----

4) 6.5
   × 0.99
   -----

5) 12.1
   × 5.2
   -----

6) 3.4
   × 8.9
   -----

7) 4.8
   × 9.1
   -----

8) 22.35
   × 20
   -----

9) 15.25
   × 3.6

✏️ Find each quotient.

10) 3.5 ÷ 0.85

11) 15.35 ÷ 4.6

12) 32.42 ÷ 8.8

13) 9.2 ÷ 3.4

14) 0.84 ÷ 0.1

15) 21.5 ÷ 1,000

16) 4.1 ÷ 100

17) 9.7 ÷ 10

18) 6.55 ÷ 1.25

19) 18.48 ÷ 11.2

MCAS Subject Test Mathematics Grade 4

Rounding Decimals

✎ Round each decimal number to the nearest place indicated.

1) 0.3<u>2</u>

2) 5.<u>0</u>1

3) 8.<u>8</u>24

4) 0.<u>4</u>78

5) <u>7</u>.32

6) 0.<u>2</u>9

7) 11.<u>3</u>1

8) <u>6</u>.223

9) 9.6<u>3</u>7

10) 5.<u>4</u>804

11) <u>7</u>.9

12) <u>5</u>.2439

13) 6.<u>4</u>92

14) 1.<u>6</u>2

15) 7<u>2</u>.85

16) 8<u>3</u>.67

17) 41.<u>6</u>8

18) 79<u>4</u>.741

19) 5<u>2</u>.2

20) 7<u>6</u>.93

21) <u>3</u>.219

22) 7<u>2</u>.09

23) 486.<u>4</u>91

24) 7.<u>0</u>8

WWW.MathNotion.Com

MCAS Subject Test Mathematics Grade 4

Comparing Decimals

✍ Write the correct comparison symbol (>, < or =).

1) 0.35 ___ 1.8

2) 1.9 ___ 1.19

3) 8.6 ___ 8.6

4) 2.45 ___ 24.5

5) 7.56 ___ 0.756

6) 11.4 ___ 11.05

7) 7.4 ___ 0.7.4

8) 8.56 ___ 0.85

9) 7 ___ 0.7

10) 7.12 ___ 0.712

11) 12.3 ___ 12.5

12) 4.67 ___ 4.68

13) 2.57 ___ 2.75

14) 3.46 ___ 0.346

15) 6.87 ___ 6.78

16) 0.89 ___ 0.98

17) 1.57 ___ 0.157

18) 0.092 ___ 0.091

19) 24.3 ___ 24.3

20) 0.17 ___ 0.71

21) 0.46 ___ 0.64

22) 0.2 ___ 0.08

23) 0.10 ___ 0.1

24) 3.52 ___ 31.5

MCAS Subject Test Mathematics Grade 4

Answers of Worksheets

Graph Decimals
1) a. 0.481 b. 0.49 c. 0.499 d. 0.502
2) a. 0.321 b. 0.326 c. 0.333 d. 0339
3) a. 0.562 b. 0.569 c. 0.577 d. 0.581
4) a. 0.414 b. 0.419 c. 0.427 d. 0431

Adding and Subtracting Decimals
1) 8.77
2) 67.65
3) 83.82
4) 32.7
5) 98.42
6) 32.56
7) 5.4
8) 6.4
9) 9.1
10) 10.7
11) 5.6
12) 5.3

Order and Comparing Decimals
1) 0.4, 0.44, 0.51, 0.67, 0.73
2) 3.9, 4.28, 4.65, 6.1, 7.02
3) 0.09, 0.72, 0.79, 1.04, 1.9
4) 1.9, 1.99, 2.6, 3.2, 4.01, 5.2
5) 2.1, 2.8, 3.8, 4.2, 5.7, 6.1
6) 0.14, 0.56, 0.87, 1.24, 3.1

Multiplying and Dividing Decimals
1) 3.15
2) 15.64
3) 15.75
4) 6.435
5) 62.92
6) 30.26
7) 43.68
8) 447
9) 54.9
10) 4.117…
11) 3.336…
12) 3.684…
13) 2.705…
14) 8.4
15) 0.0215
16) 0.041
17) 0.97
18) 5.24
19) 1.65

Rounding Decimals
1) 0.3
2) 5.0
3) 8.8
4) 0.5
5) 7
6) 0.3
7) 11.3
8) 6
9) 9.64
10) 5.5
11) 8
12) 5
13) 6.5
14) 1.6
15) 73
16) 84
17) 41.7
18) 795
19) 52
20) 77
21) 3
22) 72
23) 486.5
24) 7.1

Comparing Decimals
1) 0.35 < 1.8
2) 1.9 > 1.19
3) 8.6 = 8.6
4) 2.45 < 24.5
5) 7.56 > 0.756
6) 11.4 > 11.05
7) 7.4 > 0.74
8) 8.56 > 0.85
9) 7 > 0.7
10) 7.12 > 0.712
11) 12.3 < 12.5
12) 4.67 < 4.68

WWW.MathNotion.Com

MCAS Subject Test Mathematics Grade 4

13) 2.57 < 2.75
14) 3.46 > 0.346
15) 6.87 > 6.78
16) 0.89 < 0.98

17) 1.57 > 0.157
18) 0.092 > 0.091
19) 24.3 = 24.3
20) 0.17 < 0.71

21) 0.46 < 0.64
22) 0.2 > 0.08
23) 0.10 = 0.1
24) 3.52 < 31.5

Chapter 6 : Patterns and Algebraic Thinking

Topics that you'll learn in this chapter:

- ✓ Repeating Patterns,
- ✓ Growing Patterns,
- ✓ Patterns: Numbers,
- ✓ Finding Rules,
- ✓ Algebraic Word Problems,
- ✓ Evaluating Outputs,

MCAS Subject Test Mathematics Grade 4

Repeating Patterns

✎ Circle the picture that comes next in each picture pattern.

1)

2)

3)

4)

5)

MCAS Subject Test Mathematics Grade 4

Growing Patterns

✎ Draw the picture that comes next in each growing pattern.

1)

2)

3)

4)

5)

MCAS Subject Test Mathematics Grade 4

Patterns: Numbers

✎ Write the numbers that come next.

1) 2, 5, 8, 11,   ____, ____, ____, ____

2) 10, 15, 20, 25,   ____, ____, ____, ____

3) 4, 8, 12, 16,   ____, ____, ____, ____

4) 7, 17, 27, 37,   ____, ____, ____, ____

5) 5, 12, 19, 26,   ____, ____, ____, ____

6) 8, 16, 24, 32, 40,   ____, ____, ____, ____

✎ Write the next three numbers in each counting sequence.

7) –31, –19, –7, _____, _____, _____, _____

8) 541, 526, 511, _____, _____, _____, _____

9) 14, 34, _____, _____, 94, _____

10) 21, 29, _____, _____, _____

11) 89, 78, _____, _____, _____

12) 95, 82, 69, _____, _____, _____

13) 198, 166, 134, _____, _____, _____

14) What are the next three numbers in this counting sequence?

   1870, 1970, 2070, _____, _____, _____

15) What is the fourth number in this counting sequence?

   8, 14, 20, _____

WWW.MathNotion.Com

Finding Rules

✎ Complete the output.

1- **Rule:** the output is $x - 10.5$

Input	x	15	18	27	32.25	48.5
Output	y					

1) **Rule:** the output is $x \times 5\frac{1}{3}$

Input	x	3	9	15	21	33
Output	y					

2- **Rule:** the output is $x \div 9$

Input	x	513	387	342	198	126
Output	y					

✎ Find a rule to write an expression.

3- **Rule:** _____

Input	x	4	14	19	24
Output	y	10	35	47.5	60

4- **Rule:** _____

Input	x	5	13	19.6	34.5
Output	y	14.4	22.4	29	43.9

5- **Rule:** _____

Input	x	72	96	132	230.4
Output	y	9	12	16.5	28.8

MCAS Subject Test Mathematics Grade 4

Algebraic Word Problems

Circle the number sentence that fits the problem. Then solve for x.

1) Mary had $42. Then she earned more money (x). Now she has $86.

   $42 + x = $86   OR $42 + $86 = x

   x = ____

2) Lisa had $35. Then she earned more money (x). Now she has $78.

   $35 + x = $78   OR $35 + $78 = x

   x = ____

3) Matthew had $37. Then he earned more money (x). Now he has $98.

   $37 + x = $98   OR $37 + $98 = x

   x = ____

4) Charlotte gave 19 of the cookies he had baked to a friend and now he has 45 cookies left. 45 − 19 = x   OR   x − 19 = 45

   x = ____

5) Mia gave 32 of the cookies she had baked to a friend and now she has 55 cookies left. 55 − 32 = x   OR   x − 32 = 55

   x = ____

6) Lucas gave 41 of the cookies he had baked to a friend and now he has 49 cookies left. . 49 − 41 = x   OR   x − 41 = 49

   x = ____

MCAS Subject Test Mathematics Grade 4

Evaluating Outputs

✎ Write output of each algebraic expression.

1) $11 - x$, $x = 3$

2) $x + 14$, $x = 4$

3) $8 - 3x$, $x = 1$

4) $3x + \frac{1}{3}$, $x = \frac{1}{2}$

5) $2x + 18$, $x = 1.5$

6) $7 - 3x$, $x = 1.2$

7) $12 + 2x - 15$, $x = 3.5$

8) $25 - 5x$, $x = 2.2$

9) $\frac{44}{x} - 58$, $x = 0.4$

10) $\frac{x}{3} - 15 + x$, $x = 12.6$

11) $\frac{x}{7} + 9.5$, $x = 28.7$

12) $\frac{33}{x} - 5.2 + 2.1x$, $x = 3$

13) $2x - \frac{45}{x} - 12$, $x = 9$

14) $\frac{x}{17} - 1.8$, $x = 34$

15) $2(12.5x + 8)$, $x = 2$

16) $16x + 13x - 27 + 9$,

$x = 1.5$

17) $8.7 - \frac{16}{x} + 3x$,

$x = 4$

18) $5(3a - 2a)$,

$a = 5.5$

19) $14 - 2x + 16 - x$,

$x = 3.5$

20) $6x - 3 - x$,

$x = 1.6$

21) $18 - 2(2x + x)$, $x = 0.2$

WWW.MathNotion.Com

MCAS Subject Test Mathematics Grade 4

Answers of Worksheets

Repeating pattern

1)   2)   3)

4)   5)

Growing patterns

1)   2)   3)

4)   5)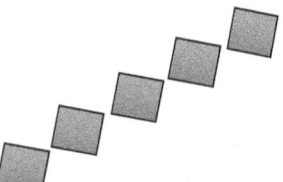

Patterns: Numbers

1) 2, 5, 8, 11, 14, 17, 20, 23

2) 10, 15, 20, 25, 30, 35, 40, 45

3) 4, 8, 12, 16, 20, 24, 28, 32

7) 5, 17, 29, 41

8) 496, 481, 466, 451

9) 14, 34, 54, 74, 94, 114

10) 37, 45, 53

11) 67, 56, 45

4) 7, 17, 27, 37, 47, 57, 67, 77

5) 5, 12, 19, 26, 33, 40, 47, 54

6) 8, 16, 24, 32, 40, 48, 56, 64

12) 56, 43, 30

13) 102, 70, 38

14) 2170, 2270, 2370

15) 26

Finding Rules

1)

Input	x	15	18	27	32.25	48.5
Output	y	4.5	7.5	16.5	21.75	38

2)

Input	x	3	9	15	21	33
Output	y	16	48	80	112	176

MCAS Subject Test Mathematics Grade 4

3)

Input	x	513	387	342	198	126
Output	y	57	43	38	22	14

4) $y = 2.5x$ 　　　　　5) $y = x + 9.4$ 　　　　　6) $y = x \div 8$

Algebraic Word Problems

1) $\$42 + x = \$86$; $x = 44$ 　　　　4) $x - 19 = 45$; $x = 64$

2) $\$35 + x = \$78$; $x = 43$ 　　　　5) $x - 32 = 55$; $x = 87$

3) $\$37 + x = \$98$; $x = 61$ 　　　　6) $x - 41 = 49$; $x = 90$

Evaluating Outputs

1) 8　　　　　　7) 4　　　　　　13) 1　　　　　　19) 19.5

2) 18　　　　　8) 14　　　　　14) 0.2　　　　　20) 5

3) 5　　　　　　9) 52　　　　　15) 66　　　　　21) 16.8

4) $\frac{11}{6}$　　　　10) 1.8　　　　16) 25.5

5) 21　　　　　11) 13.6　　　　17) 16.7

6) 3.4　　　　12) 12.1　　　　18) 27.5

Chapter 7 : Measurement

Topics that you'll learn in this chapter:

- ✓ Reference Measurement Units,
- ✓ Metric Length Units,
- ✓ Customary Length Units,
- ✓ Metric Capacity Units,
- ✓ Customary Capacity Units,
- ✓ Metric Weight and Mass Units,
- ✓ Customary Weight and Mass Units,
- ✓ Temperature Units,
- ✓ Time,
- ✓ Add Money Amounts,
- ✓ Subtract Money Amounts,
- ✓ Money: Word Problems,

MCAS Subject Test Mathematics Grade 4

Reference Measurement Units

LENGTH

Customary

1 mile (mi) = 1,760 yards (yd)

1 yard (yd) = 3 feet (ft)

1 foot (ft) = 12 inches (in.)

Metric

1 kilometer (km) = 1,000 meters (m)

1 meter (m) = 100 centimeters (cm)

1 centimeter (cm) = 10 millimeters (mm)

VOLUME AND CAPACITY

Customary

1 gallon (gal) = 4 quarts (qt)

1 quart (qt) = 2 pints (pt.)

1 pint (pt.) = 2 cups (c)

1 cup (c) = 8 fluid ounces (Fl oz)

Metric

1 liter (L) = 1,000 milliliters (mL)

WEIGHT AND MASS

Customary

1 ton (T) = 2,000 pounds (lb.)

1 pound (lb.) = 16 ounces (oz)

Metric

1 kilogram (kg) = 1,000 grams (g)

1 gram (g) = 1,000 milligrams (mg)

Time

1 year = 12 months

1 year = 52 weeks

1 week = 7 days

1 day = 24 hours

1 hour = 60 minutes

1 minute = 60 seconds

WWW.MathNotion.Com

MCAS Subject Test Mathematics Grade 4

Metric Length Units

✏ Convert to the units.

1) 300 mm = _____ cm

2) 8 m = _____ mm

3) 4.5 m = _____ cm

4) 7 km = _____ m

5) 9,400 mm = _____ m

6) 1,100 cm = _____ m

7) 2.8 m = _____ cm

8) 4,000 mm = _____ cm

9) 7,000 mm = _____ m

10) 2 km = _____ mm

11) 14.9 km = _____ m

12) 20 m = _____ cm

13) 5,000 m = _____ km

14) 7,600 m = _____ km

Customary Length Units

✏ Convert to the units.

1) 8 ft = _____ in

2) 4 ft = _____ in

3) 6 yd = _____ ft

4) 10 yd = _____ ft

5) 3,520 yd = _____ mi

6) 60 in = _____ ft

7) 144 in = _____ yd

8) 0.5 mi = _____ yd

9) 15 yd = _____ in

10) 42 yd = _____ in

11) 99 ft = _____ yd

12) 1.5 mi = _____ yd

13) 84 in = _____ ft

14) 30 yd = _____ feet

WWW.MathNotion.Com

MCAS Subject Test Mathematics Grade 4

Metric Capacity Units

✎ Convert the following measurements.

1) 32.4 l = _____ ml

2) 7.1 l = _____ ml

3) 54 l = _____ ml

4) 92 l = _____ ml

5) 48 l = _____ ml

6) 13 l = _____ ml

7) 750 ml = _____ l

8) 2,400 ml = _____ l

9) 73,000 ml = _____ l

10) 8,000 ml = _____ l

11) 49,000 ml = _____ l

12) 5,500 ml = _____ l

Customary Capacity Units

✎ Convert the following measurements.

1) 51 gal = _____ qt.

2) 35 gal = _____ pt.

3) 68 gal = _____ c.

4) 20 pt. = _____ c

5) 12.5 qt = _____ pt.

6) 22.5 qt = _____ c

7) 51 pt. = _____ c

8) 48 c = _____ gal

9) 96 pt. = _____ gal

10) 136 qt = _____ gal

11) 15 c = _____ fl oz

12) 44 c = _____ qt

13) 240 c = _____ pt.

14) 148 qt = _____ gal

15) 160 pt. = _____ qt

16) 104 fl oz = _____ c.

WWW.MathNotion.Com

MCAS Subject Test Mathematics Grade 4

Metric Weight and Mass Units

✍ Convert.

1) 60 kg = _____ g

2) 24 kg = _____ g

3) 610 kg = _____ g

4) 82 kg = _____ g

5) 95.8 kg = _____ g

6) 4.85 kg = _____ g

7) 1.5 kg = _____ g

8) 95,000 g = _____ kg

9) 241,000 g = _____ kg

10) 700,000 g = _____ kg

11) 2,500 g = _____ kg

12) 28,900 g = _____ kg

13) 970,000 g = _____ kg

14) 325,500 g = _____ kg

Customary Weight and Mass Units

✍ Convert.

1) 10,000 lb. = _____ T

2) 19,000 lb. = _____ T

3) 32,000 lb. = _____ T

4) 16,800 lb. = _____ T

5) 27 lb. = _____ oz

6) 25.4 lb. = _____ oz

7) 124 lb. = _____ oz

8) 4T = _____ lb.

9) 7T = _____ lb.

10) 11.2T = _____ lb.

11) 12.8T = _____ lb.

12) $\frac{6}{5}$ T = _____ oz

13) 9.125 T = _____ oz

14) $\frac{3}{4}$ T = _____ oz

MCAS Subject Test Mathematics Grade 4

Time

✏️ Convert to the units.

1) 22 hr. = _____ min

2) 12.5 year = _____ week

3) 5.2 hr = _____ sec

4) 21 min = _____ sec

5) 1,800 min = _____ hr.

6) 730 day = _____ year

7) 1.5 year = _____ hr.

8) 42 day = _____ hr.

9) 2 day = _____ min

10) 660 min = _____ hr.

11) 10 year = _____ month

12) 2,124 sec = _____ min

13) 168 hr = _____ day

14) 19 weeks = _____ day

✏️ How much time has passed?

1) From 2:25 A.M. to 5:35 A.M.: ____ hours and ___ minutes.

2) From 3:30 A.M. to 7:55 A.M.: ____ hours and ___ minutes.

3) It's 6:30 P.M. What time was 2 hours ago? _____ O'clock

4) 4:30 A.M to 7:50 AM: _____ hours and _____ minutes.

5) 4:15 A.M to 8:35 AM: _____ hours and _____ minutes.

6) 5:10 A.M. to 7:35 AM. = _____ hour(s) and _____ minutes.

7) 10:55 A.M. to 3:25 PM. = _____ hour(s) and _____ minutes

8) 8:05 A.M. to 8:40 A.M. = _____ minutes

9) 6:02 A.M. to 6:49 A.M. = _____ minutes

MCAS Subject Test Mathematics Grade 4

Money Amounts

🖎 Add.

1) $128
   +$328
   ———

   $225
   +$245
   ———

   $150
   +$186
   ———

2) $453
   +$128
   ———

   $440
   +$541
   ———

   $258
   +$248
   ———

3) $645
   +$112.5
   ———

   $235.4
   +$452.1
   ———

   $125.99
   +$148.32
   ———

4) $321.40
   +$175.80
   ———

   $458.10
   +$752.65
   ———

   $652.00
   +$324.70
   ———

🖎 Subtract.

5) $725
   −$334
   ———

   $543
   −$248
   ———

   $349
   −$122
   ———

6) $658.20
   −$220.30
   ———

   $752.10
   −$452.15
   ———

   $312.50
   −$89.90
   ———

7) $315.90
   −$220.10
   ———

   $548.40
   −$342.10
   ———

   $968.40
   −$324.50
   ———

8) Linda had $18.60. She bought some game tickets for $9.25. How much did she have left?

MCAS Subject Test Mathematics Grade 4

Money: Word Problems

✎ Solve.

1) How many boxes of envelopes can you buy with $40 if one box costs $8?

2) After paying $7.22 for a salad, Ella has $45.86. How much money did she have before buying the salad?

3) How many packages of diapers can you buy with $96 if one package costs $6?

4) Last week James ran 28.5 miles more than Michael. James ran 59 miles. How many miles did Michael run?

5) Last Friday Jacob had $14.68. Over the weekend he received some money for cleaning the attic. He now has $38.95. How much money did he receive?

6) After paying $3.15 for a sandwich, Amelia has $48.69. How much money did she have before buying the sandwich?

MCAS Subject Test Mathematics Grade 4

Answers of Worksheets

Metric length

1) 30 cm
2) 8,000 mm
3) 450 cm
4) 7,000 m
5) 9.4 m
6) 11 m
7) 280 cm
8) 400 cm
9) 7 m
10) 2,000,000 mm
11) 14,900 m
12) 2,000 cm
13) 5 km
14) 7.6 km

Customary Length

1) 96
2) 48
3) 18
4) 30
5) 2
6) 5
7) 4
8) 880
9) 540
10) 1,512
11) 33
12) 2,640
13) 7
14) 90

Metric Capacity

1) 32,400 ml
2) 7,100 ml
3) 54,000 ml
4) 92,000 ml
5) 48,000 ml
6) 13,000 ml
7) 0.75ml
8) 2.4 ml
9) 73 ml
10) 8L
11) 49 L
12) 5.5 L

Customary Capacity

1) 204 qt
2) 280 pt.
3) 1,088 c
4) 40 c
5) 25 pt.
6) 90c
7) 102 c
8) 3 gal
9) 12 gal
10) 34 gal
11) 120 qt
12) 11qt
13) 120 pt.
14) 37 gal
15) 80 qt
16) 13 pt.

Metric Weight and Mass

1) 60,000 g
2) 24,000 g
3) 610,000 g
4) 82,000 g
5) 95,800g
6) 4,850 g
7) 1,500 g
8) 95 kg
9) 241 kg
10) 700 kg
11) 2.5 kg
12) 28.9 kg
13) 970 kg
14) 325.5 kg

Customary Weight and Mass

1) 5 T
2) 9.5 T
3) 16 T
4) 8.4 T
5) 432 oz
6) 406.4 oz

WWW.MathNotion.Com

MCAS Subject Test Mathematics Grade 4

7) 1,984 oz
8) 8,000 lb.
9) 14,000 lb.
10) 22,400 lb.
11) 25,600 lb.
12) 38,400 oz
13) 292,000 oz
14) 24,000 oz

Time - Convert

1) 1,320 min
2) 650 weeks
3) 18,720 sec
4) 1,260 sec
5) 30 hr
6) 2 year
7) 13,140 hr
8) 1,008 hr
9) 2,880 min
10) 11hr
11) 120 months
12) 35.4 min
13) 7 days
14) 133 days

Time - Gap

1) 3:10
2) 4:25
3) 4:30P.M.
4) 3:20
5) 4:20
6) 2:25
7) 4:30
8) 35 minutes
9) 47 minutes

Add Money

1) 456, 470, 336
2) 581, 981, 506
3) 757.5, 687.5, 274.31
4) 497.2, 1210.75, 976.7

Subtract Money

5) 391, 295, 227
6) 437.9, 299.95, 222.6
7) 95.8, 206.3, 643.9
8) $9.35

Money: word problem

1) 5
2) $53.08
3) 16
4) 30.5
5) 24.27
6) 51.84

Chapter 8 : Geometric

Topics that you'll learn in this chapter:

- ✓ Identifying Angles,
- ✓ Estimate and Measure Angles,
- ✓ Polygon Names,
- ✓ Classify Triangles,
- ✓ Parallel Sides in Quadrilaterals,
- ✓ Identify Parallelograms,
- ✓ Identify Trapezoids,
- ✓ Identify Rectangles,
- ✓ Perimeter and Area of Squares,
- ✓ Perimeter and Area of rectangles,
- ✓ Area and Perimeter: Word Problems,
- ✓ Volume,

MCAS Subject Test Mathematics Grade 4

Identifying Angles

✏️ Write the name of the angles (Acute, Right, Obtuse, and Straight).

1)

2)

3)

4)

5)

6)

7)

8)

WWW.MathNotion.Com

MCAS Subject Test Mathematics Grade 4

Estimate Angle Measurements

✏ Estimate the approximate measurement of each angle in degrees.

1)

2)

3)

4)

5)

6)

7)

8)

MCAS Subject Test Mathematics Grade 4

Measure Angles with a Protractor

✏️ Use protractor to measure the angles below.

1)

2)

3)

4)

✏️ Use a protractor to draw angles for each measurement given.

1) 140°

2) 100°

3) 110°

4) 120°

5) 55°

Polygon Names

✎ Write name of polygons.

1)

2)

3)

4)

5)

6)

Classify Triangles

✏️ Classify the triangles by their sides and angles.

1)

2)

3)

4)

5)

6)

Parallel Sides in Quadrilaterals

✎ Write name of quadrilaterals.

1)

2)

3)

4)

5)

6)

MCAS Subject Test Mathematics Grade 4

Identify Rectangles

✏ Solve.

1) A rectangle has _____ sides and _____ angles.

2) Draw a rectangle that is 5.5 centimeters long and 2.5 centimeters wide. What is the perimeter?

3) Draw a rectangle 3.5 cm long and 1.5 cm wide.

4) Draw a rectangle whose length is 4.25 cm and whose width is 2.45 cm. What is the perimeter of the rectangle?

5) What is the perimeter of the rectangle?

7.2

5.8

MCAS Subject Test Mathematics Grade 4

Perimeter: Find the Missing Side Lengths

✏ Find the missing side of each shape.

1) perimeter = 57.2

2) perimeter = 21.2

3) perimeter = 27.5

4) perimeter = 35.2

5) perimeter = 75.6

6) perimeter = 30.8

7) perimeter = 36.25

8) perimeter = 46.8

WWW.MathNotion.Com

MCAS Subject Test Mathematics Grade 4

Perimeter and Area of Squares

✏️ Find perimeter and area of squares.

1) A: _____ , P: _____

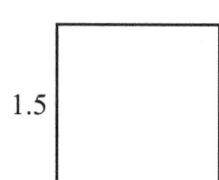

2) A: _____ , P: _____

3) A: _____ , P: _____

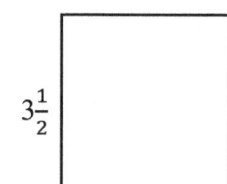

4) A: _____ , P: _____

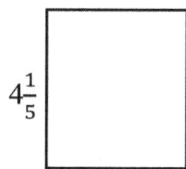

5) A: _____ , P: _____

6) A: _____ , P: _____

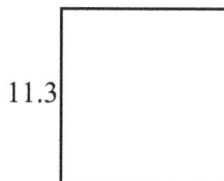

7) A: _____ , P: _____

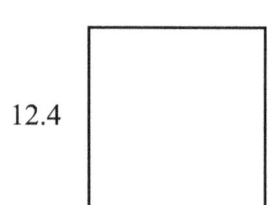

8) A: _____ , P: _____

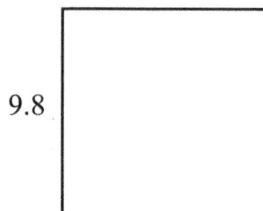

MCAS Subject Test Mathematics Grade 4

Perimeter and Area of rectangles

✎ Find perimeter and area of rectangles.

1) A: _____ , P: _____

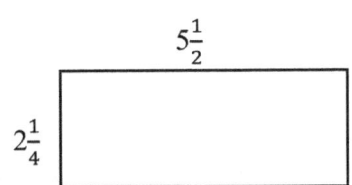

2) A: _____ , P: _____

3) A: _____ , P: _____

6.1
8.3

4) A: _____ , P: _____

11.9
10.6

5) A: _____ , P: _____

7.6
$5\frac{1}{4}$

6) A: _____ , P: _____

7) A: _____ , P: _____

9.7
$3\frac{1}{2}$

8) A: _____ , P: _____

11.4
5.8

WWW.MathNotion.Com   91

MCAS Subject Test Mathematics Grade 4

Find the Area or Missing Side Length of a Rectangle

🖊 Find area or missing side length of rectangles.

1) Area =?

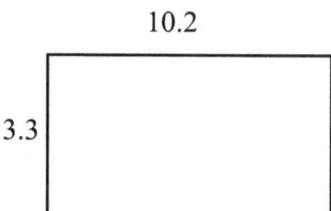
10.2
3.3

2) Area = 42.12, x=?

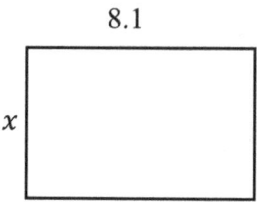
8.1
x

3) Area = 29.52, x=?

4.1
x

4) Area =?

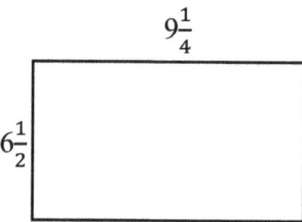
$9\frac{1}{4}$
$6\frac{1}{2}$

5) Area =?

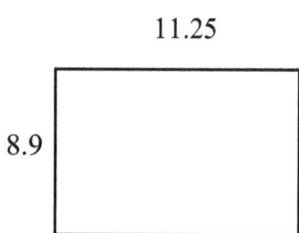
11.25
8.9

6) Area = 662.34 x=?

23.24
x

7) Area = 216.24, x=?

$21\frac{1}{5}$
x

8) Area 336.42, x=?

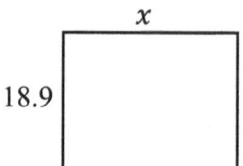
x
18.9

Area and Perimeter: Word Problems

✎ Solve.

1) The area of a rectangle is 90.86 square meters. The width is 7.7 meters. What is the length of the rectangle?

2) A square has an area of 6.25 square feet. What is the perimeter of the square?

3) Ava built a rectangular vegetable garden that is 3.2 feet long and has an area of 21.12 square feet. What is the perimeter of Ava's vegetable garden?

4) A square has a perimeter of 12.8 millimeters. What is the area of the square?

5) The perimeter of David's square backyard is 0.96 meters. What is the area of David's backyard?

6) The area of a rectangle is 37.63 square inches. The length is 7.1 inches. What is the perimeter of the rectangle?

MCAS Subject Test Mathematics Grade 4

Volume of Cubes and Rectangle Prisms

🖎 Find the volume of each of the rectangular prisms.

1)

2)

3)

4)

5)

6)

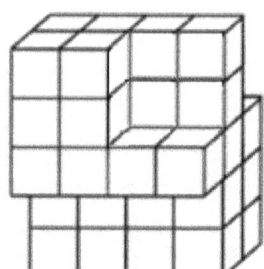

WWW.MathNotion.Com

MCAS Subject Test Mathematics Grade 4

Answers of Worksheets

Identifying Angles

1) Right   3) Obtuse   5) Acute   7) Obtuse
2) Acute   4) Straight   6) Obtuse   8) Acute

Estimate Angle Measurements

1) 160°   3) 110°   5) 130°   7) 90°
2) 180°   4) 75°   6) 45°   8) 60°

Measure Angles with a Protractor

1) 50°   2) 135°   3) 20°   4) 170°

Draw Angles

1)   2)   3)

4)   5)

 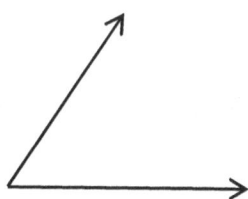

Polygon Names

1) Diamond   3) Pentagon   5) Heptagon
2) Parallelogram   4) Trapezius   6) Octagon

Classify Triangles

1) Scalene, acute   4) Scalene, right
2) Isosceles, acute   5) Isosceles, right
3) Equilateral, acute   6) Scalene, obtuse

MCAS Subject Test Mathematics Grade 4

Parallel Sides in Quadrilaterals

1) Hexagon
2) Kike
3) Parallelogram
4) Trapezoid
5) Rhombus
6) Rectangle

Identify Rectangles

1) 4 - 4
2) 16
3) Draw the rectangle.
4) 13.4
5) 26

Perimeter: Find the Missing Side Lengths

1) 14.3
2) 7.2
3) 5.5
4) 8.8
5) 18.9
6) 6.2
7) 7
8) 11.7

Perimeter and Area of Squares

1) A: 2.25, P: 6
2) A: 27.56, P: 21
3) A: 12.25, P: 14
4) A: 17.64, P: 16.8
5) A: 105.063 P: 41
6) A: 127.69, P: 45.2
7) A: 153.76, P: 49.6
8) A: 96.04, P: 39.2

Perimeter and Area of rectangles

1) A: 12.375, P: 15.5
2) A: 27.3, P: 21.4
3) A: 50.63, P: 28.8
4) A: 126.14, P: 45
5) A: 39.9, P: 25.7
6) A: 26.46, P: 21
7) A: 33.95, P: 26.4
8) A: 66.12, P: 34.4

Find the Area or Missing Side Length of a Rectangle

1) 33.66
2) 5.2
3) 7.2
4) 60.125
5) 100.125
6) 28.5
7) 10.2
8) 17.8

Area and Perimeter: Word Problems

1) 11.8
2) 10
3) 19.6
4) 10.24
5) 0.0576
6) 24.8

Volume of Cubes and Rectangle Prisms

1) 452.025 cm^3
2) 429.2 cm^3
3) 148.877 c m^3
4) 583.848 cm^3
5) 32
6) 40

MCAS Subject Test Mathematics Grade 4

Chapter 9 : Data Graphs, and Statistics

Topics that you'll learn in this chapter:

- ✓ Bar Graph,
- ✓ Tally and Pictographs,
- ✓ Dot Plots
- ✓ Line Graphs,
- ✓ Stem–And–Leaf Plot,
- ✓ Coordinate Plane,

Bar Graph

✎ Graph the given information as a bar graph.

Day	Hot dogs sold
Monday	50
Tuesday	80
Wednesday	10
Thursday	30
Friday	70

MCAS Subject Test Mathematics Grade 4

Tally and Pictographs

🖉 Using the key, draw the pictograph to show the information.

Key: ⚽ = 2 animals

MCAS Subject Test Mathematics Grade 4

Dot plots

The ages of students in a Math class are given below.

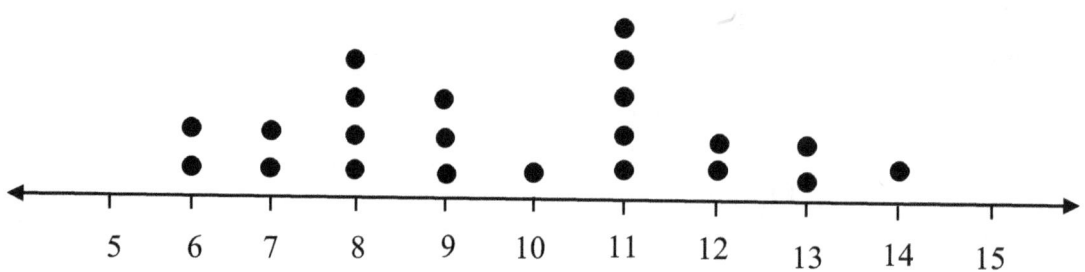

1) What is the total number of students in math class?

2) How many students are at least 12 years old?

3) Which age(s) has the most students?

4) Which age(s) has the fewest student?

5) Determine the median of the data.

6) Determine the range of the data.

7) Determine the mode of the data.

WWW.MathNotion.Com

MCAS Subject Test Mathematics Grade 4

Line Graphs

David work as a salesman in a store. He records the number of shoes sold in five days on a line graph. Use the graph to answer the question.

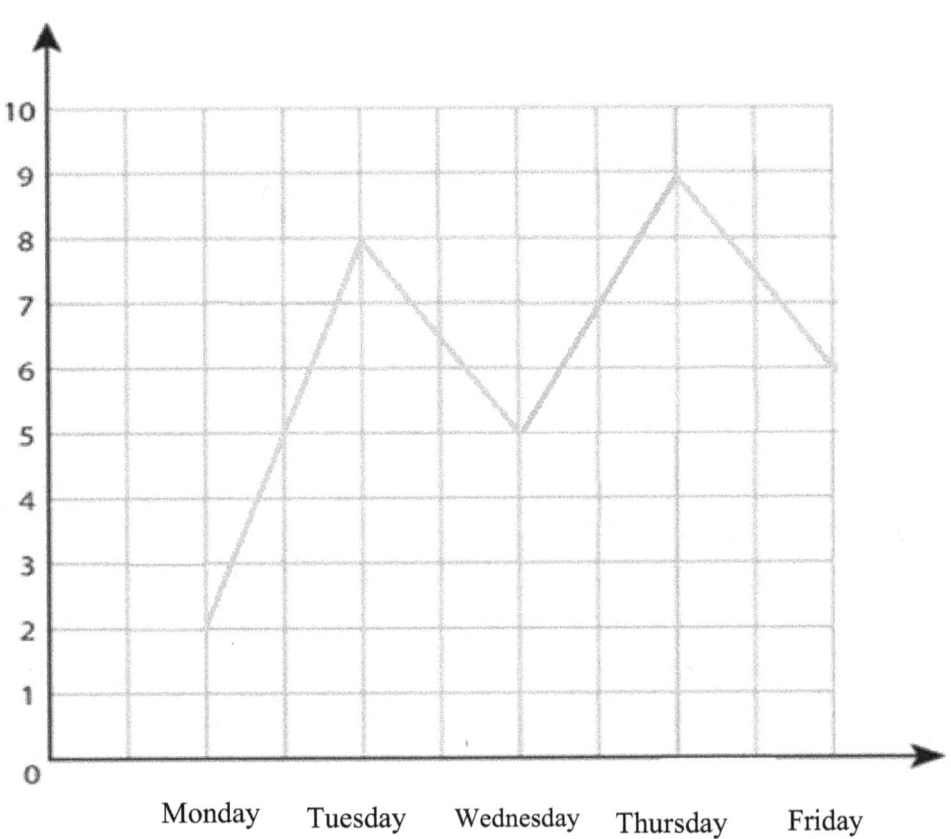

1) How many shoes were sold on Friday?

2) Which day had the minimum sales of shoes?

3) Which day had the maximum number of shoes sold?

4) How many shoes were sold in 5 days?

Stem–And–Leaf Plot

✎ Make stem ad leaf plots for the given data.

1) 42, 47, 14, 19, 42, 69, 65, 49, 42, 10, 64

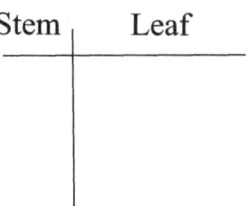

2) 43, 85, 52, 48, 45, 43, 51, 81, 59, 50, 85, 89

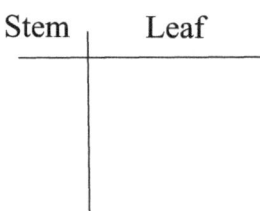

3) 112, 39, 46, 35, 80, 119, 42, 114, 37, 112, 47, 119

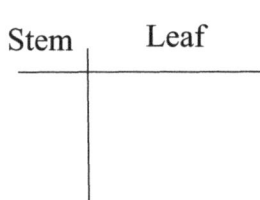

4) 90, 50, 131, 93, 112, 56, 139, 98, 115, 59, 98, 135, 111

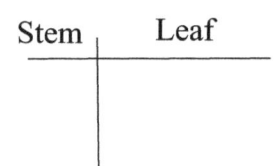

Coordinate Plane

 Plot each point on the coordinate grid.

1) A (4, 6)   3) C (1, 5)   5) E (4, 8)
2) B (3, 2)   4) D (5, 7)   6) F (9, 2)

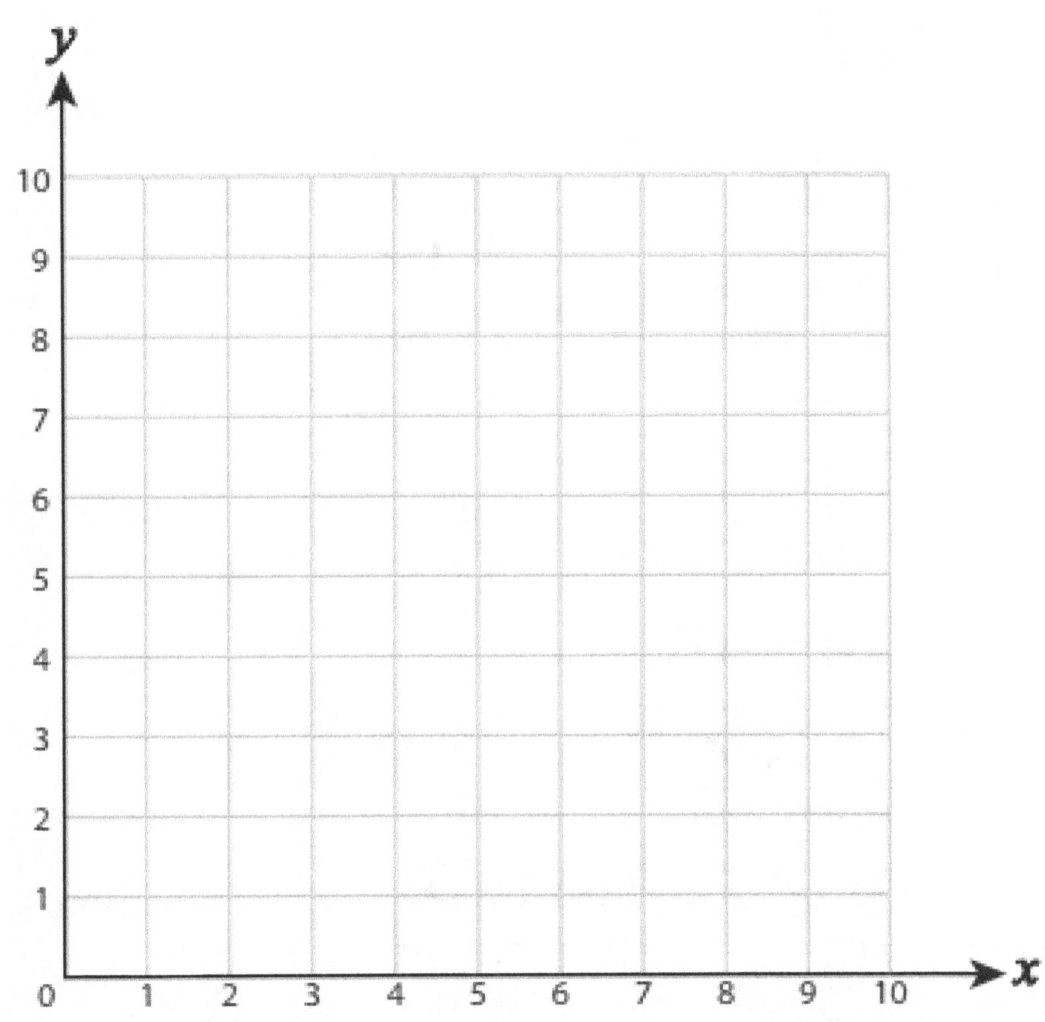

MCAS Subject Test Mathematics Grade 4

Answers of Worksheets

Bar Graph

Tally and Pictographs

Dot plots

1) 22
2) 5
3) 11
4) 10 and 14
5) 2
6) 4
7) 2

MCAS Subject Test Mathematics Grade 4

Line Graphs

1) 6 2) Monday 3) Thursday 4) 30

Stem–And–Leaf Plot

1)
Stem	leaf
1	0 4 9
4	2 2 2 7 9
6	4 5 9

2)
Stem	leaf
4	3 3 5 8
5	0 1 2 9
8	1 5 5 9

3)
Stem	leaf
3	5 7 9
4	2 6 7
8	0
11	2 2 4 9 9

4)
Stem	leaf
5	0 6 9
9	0 3 8 8
11	1 2 5
13	1 5 9

Coordinate Plane

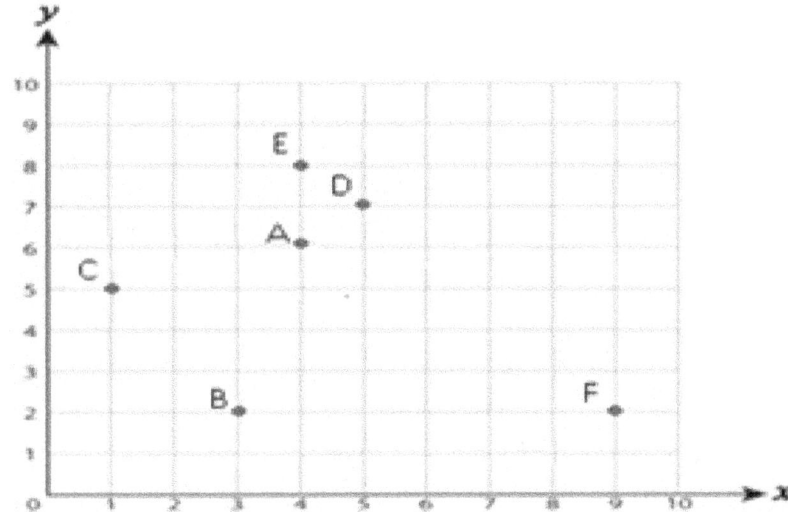

MCAS Subject Test Mathematics Grade 4

Chapter 10 : Three-Dimensional Figures

Topics that you'll learn in this chapter:

- ✓ Identify Three–Dimensional Figures,
- ✓ Count Vertices, Edges, and Faces,
- ✓ Identify Faces of Three–Dimensional Figures,

Identify Three-Dimensional Figures

✎ Write the name of each shape.

1)

2)

3)

4)

5)

6)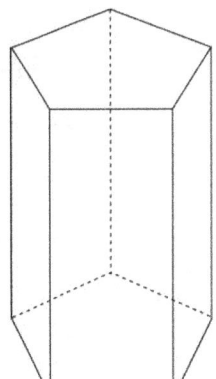

Count Vertices, Edges, and Faces

	Shape	Number of edges	Number of faces	Number of vertices
1)		___	___	___
2)		___	___	___
3)		___	___	___
4)		___	___	___
5)		___	___	___
6)		___	___	___

MCAS Subject Test Mathematics Grade 4

Identify Faces of Three–Dimensional Figures

✎ Write the number of faces.

1)

2)

3)

4)

5)

6)

7)

8)

WWW.MathNotion.Com

MCAS Subject Test Mathematics Grade 4

Answers of Worksheets

Identify Three–Dimensional Figures

1) Cube
2) Triangular pyramid
3) Triangular prism
4) Square pyramid
5) Rectangular prism
6) Pentagonal prism
7) Hexagonal prism

Count Vertices, Edges, and Faces

Shape	Number of edges	Number of faces	Number of vertices
1)	6	4	4
2)	8	5	5
3)	12	6	8
4)	15	7	10
5)	12	6	8
6)	18	8	12

Identify Faces of Three–Dimensional Figures

1) 6
2) 2
3) 5
4) 4
5) 6
6) 7
7) 8
8) 5

WWW.MathNotion.Com

MCAS Subject Test Mathematics Grade 4

Chapter 11 : Symmetry and Transformations

Topics that you'll learn in this chapter:

- ✓ Line Segments,
- ✓ Identify Lines of Symmetry,
- ✓ Count Lines of Symmetry,
- ✓ Parallel, Perpendicular and Intersecting Lines,

MCAS Subject Test Mathematics Grade 4

Line Segments

✎ Write each as a line, ray, or line segment.

1)

2)

3)

4)

5)

6)

7)

8)

WWW.MathNotion.Com

MCAS Subject Test Mathematics Grade 4

Identify Lines of Symmetry

Tell whether the line on each shape a line of symmetry is.

1)

2)

3)

4)

5)

6)

7)

8)
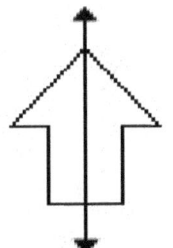

WWW.MathNotion.Com

MCAS Subject Test Mathematics Grade 4

Count Lines of Symmetry

✎Draw lines of symmetry on each shape. Count and write the lines of symmetry you see.

1)

2)

3)

4)

5)

6)

7)

8)

WWW.MathNotion.Com

MCAS Subject Test Mathematics Grade 4

Parallel, Perpendicular and Intersecting Lines

✏️ State whether the given pair of lines are parallel, perpendicular, or intersecting.

1)

2)

3)

4)

5)

6)

7)

8)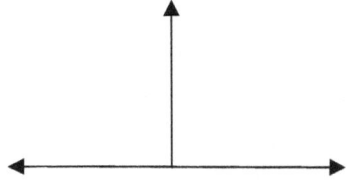

MCAS Subject Test Mathematics Grade 4

Answers of Worksheets

Line Segments

1) Ray
2) Line segment
3) Line
4) Ray
5) Ray
6) Line
7) Line
8) Line segment

Identify lines of symmetry

1) yes
2) no
3) no
4) yes
5) yes
6) yes
7) no
8) yes

Count lines of symmetry

1)

2)

3)

4)

5)

6)

7)

8)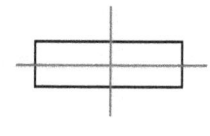

Parallel, Perpendicular and Intersecting Lines

1) Intersection
2) Perpendicular
3) Parallel
4) Intersection
5) Intersection
6) Perpendicular
7) Parallel
8) Perpendicular

MCAS Subject Test Mathematics Grade 4

Chapter 12 : MCAS Math Practice Tests

Time to Test

Time to refine your skill with a practice examination.

Take a REAL MCAS Mathematics test to simulate the test day experience. After you have finished, score your test using the answer key.

Before You Start

- You will need a pencil and scratch papers to take the test.
- For this practice test, do not time yourself. Spend time as much as you need.
- It is okay to guess. You will not lose any points if you are wrong.
- After you have finished the test, review the answer key to see where you went wrong.

Calculators are not permitted for Grade 4 MCAS Tests

Good Luck!

MCAS Subject Test Mathematics Grade 4

MCAS GRADE 4 MAHEMATICS REFRENCE MATERIALS

LENGTH

Customary

1 mile (mi) = 1,760 yards (yd)

1 yard (yd) = 3 feet (ft)

1 foot (ft) = 12 inches (in.)

Metric

1 kilometer (km) = 1,000 meters (m)

1 meter (m) = 100 centimeters (cm)

1 centimeter (cm) = 10 millimeters (mm)

VOLUME AND CAPACITY

Customary

1 gallon (gal) = 4 quarts (qt)

1 quart (qt) = 2 pints (pt.)

1 pint (pt.) = 2 cups (c)

1 cup (c) = 8 fluid ounces (Fl oz)

Metric

1 liter (L) = 1,000 milliliters (mL)

WEIGHT AND MASS

Customary

1 ton (T) = 2,000 pounds (lb.)

1 pound (lb.) = 16 ounces (oz)

Metric

1 kilogram (kg) = 1,000 grams (g)

1 gram (g) = 1,000 milligrams (mg)

Time

1 year = 12 months

1 year = 52 weeks

1 week = 7 days

1 day = 24 hours

1 hour = 60 minutes

1 minute = 60 seconds

Perimeter

Square $P = 4S$

Rectangle $P = L + W + L + W$ or $P = 2L + 2W$

Area

Square $A = S \times S$

Rectangle $A = L \times W$

Massachusetts Comprehensive Assessment System Practice Test 1

Mathematics

GRADE 4

Released

MCAS Subject Test Mathematics Grade 4

Session 1

❖ **Calculators are NOT permitted for this practice test.**

❖ **Time for Session 1: 85 Minutes**

MCAS Subject Test Mathematics Grade 4

1) If this clock shows a time in the morning, what time was it 3 hours and 20 minutes ago?

   A. 09:20 AM

   B. 07:55 AM

   C. 08:05 AM

   D. 07:05 PM

2) There are 7 days in a week. There are 31 days in the month of March. How many times as many days are there in March than are in one week?

   Write your answer in the box below.

   ☐

3) A football team is buying new uniforms. Each uniform cost $16. The team wants to buy 12 uniforms.

   Which equation represents a way to find the total cost of the uniforms?

   A. $(16 \times 12) + (16 \times 2) = 192 + 32$

   B. $(16 \times 6) + (12 \times 6) = 96 + 72$

   C. $(16 \times 10) + (16 \times 2) = 160 + 32$

   D. $(16 \times 5) + (5 \times 12) = 90 + 62$

MCAS Subject Test Mathematics Grade 4

4) Use the table below to answer the question.

Favorite Sports

Sport	Number of Votes
football (FB)	11
basketball (BB)	2
soccer (SOC)	7
volleyball (VB)	6

The students in the fourth-grade class voted for their favorite sport. Which bar graph shows results of the students vote?

A.

B.

C.

D.
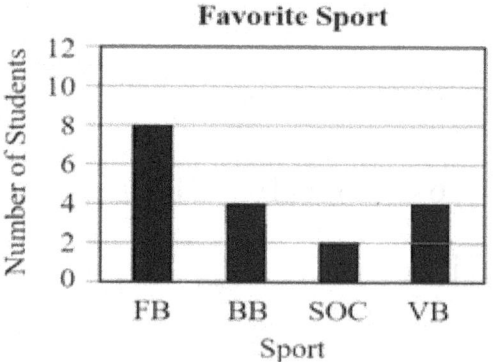

MCAS Subject Test Mathematics Grade 4

5) A number sentence such as $32 + Z = 85$ can be called an equation. If this equation is true, then which of the following equations is not true?

A. $85 - 32 = Z$

B. $85 - Z = 32$

C. $Z - 32 = 85$

D. $Z + 32 = 85$

6) Tam has 720 cards. He wants to put them in boxes of 80 cards. How many boxes does he need?

Write your answer in the box below.

☐

7) Use the table below to answer the question.

Which list of city populations is in order from least to greatest?

A. 23,031; 12,850; 28,017; 21,080.

B. 12,850; 21,080; 23,031; 28,017.

C. 21,080; 23,031; 28,017; 12,850.

D. 28,017; 23,031; 21,080; 12,850.

City Populations	
City	Population
Denton	12,850
Bamberg	21,080
Windham	23,031
Sand hill	28,017

MCAS Subject Test Mathematics Grade 4

8) Which number correctly completes the subtraction sentence 8.0 – 2.55 = _____?

   A. 6.45

   B. 5.45

   C. 4.75

   D. 5.35

9) For a concert, there are children's tickets and adult tickets for sale. Of the total available tickets, $\frac{18}{100}$ have been sold as adult tickets and $\frac{7}{10}$ as children's tickets.

   The rest of the tickets have not been sold.

   What fraction of the total number of tickets for the concert have been sold?

   A. $1\frac{11}{25}$

   B. $\frac{22}{25}$

   C. $\frac{25}{100}$

   D. $\frac{60}{100}$

10) Circle a reasonable measurement for the angle F:

    A. 40°

    B. 95°

    C. 100°

    D. 240°

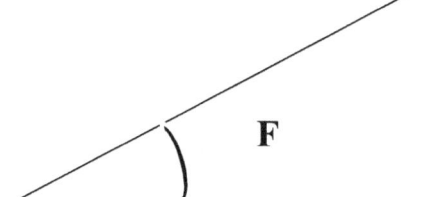

MCAS Subject Test Mathematics Grade 4

Session 2

❖ Calculators are NOT permitted for this practice test.

❖ Time for Session 2: 85 Minutes

MCAS Subject Test Mathematics Grade 4

11) Mia has a group of shapes. Each shape in her group has at least one set of parallel sides. Each shape also has at least one set of perpendicular sides. Which group could be Mia's group of shapes?

A.

B.

C.

D.

12) A stack of 4 pennies has a height of 1 centimeter. Elise has a stack of pennies with a height of 9 centimeters. Which equation can be used to find the number of pennies, n, in Elise's stack of pennies?

A. $n = 9 + 4$

B. $n = 9 - 4$

C. $n = 4 \times 9$

D. $n = 9 \div 4$

MCAS Subject Test Mathematics Grade 4

13) Use the models below to answer the question.

Which statement about the models is true?

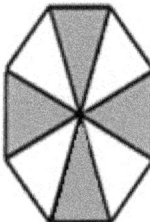

A. Each shows the same fraction because they are the same size.

B. Each shows a different fraction because they are different shapes.

C. Each shows the same fraction because they both have 4 sections shaded.

D. Each shows a different fraction because they both have 4 shaded sections but a different number of total sections.

14) What is the value of A in the equation $45 \div A = 5$?

Write your answer in the box below.

15) Write $\frac{77}{100}$ as a decimal number.

A. 7.77

B. 0.77

C. 77.7

D. 0.077

MCAS Subject Test Mathematics Grade 4

16) Erik made 56 pints of juice. He drinks 7 cups of juice each day. How many days will Erik take to drink all of the juice he made?

   A. 6 days

   B. 8 days

   C. 9 days

   D. 12 days

17) Emma described a number using these clues.

   ✓ 3 digits of the number are 3, 5, and 9.

   ✓ The value of the digit 5 is (5 × 10).

   ✓ The value of the digit 9 is (9 × 10,000).

   ✓ The value of the digit 3 is (3 × 1,000).

   Which number could fit Emma's description?

   A. 64,935.71

   B. 93,465.70

   C. 93,654.71

   D. 39,654.71

18) There are 24 boxes and each box contain 14 pencils. How many pencils are in the boxed in total?

   A. 166

   B. 266

   C. 336

   D. Not here

MCAS Subject Test Mathematics Grade 4

19) Moe packs 20 boxes with flashcards. Each box holds 80 flashcards. How many flashcards Moe can pack into these boxes?

A. 1,600

B. 3,200

C. 160

D. 1,800

20) Sophia flew 6,708 miles from Los Angeles to New York City. What is the number of miles Sophia flew rounded to the nearest thousand?

Write your answer in the box below.

☐

"This is the end of Practice Test 1."

MCAS Subject Test Mathematics Grade 4

Massachusetts Comprehensive Assessment System Practice Test 2

Mathematics

GRADE 4

Released Month Year

Session 1

- ❖ Calculators are NOT permitted for this practice test.
- ❖ Time for Session 1: 85 Minutes

MCAS Subject Test Mathematics Grade 4

1) Jeb paid $168 for a magazine subscription. If he is paying $6 for each issue of the magazine, how many issues of the magazine will he receive?

   A. 28

   B. 26

   C. 30

   D. 32

2) A straight-line measure 180°. A straight line and a triangle are touching as shown in the figure below.

   What is the value of A in the figure?

   A. 94

   B. 82

   C. 84

   D. 92

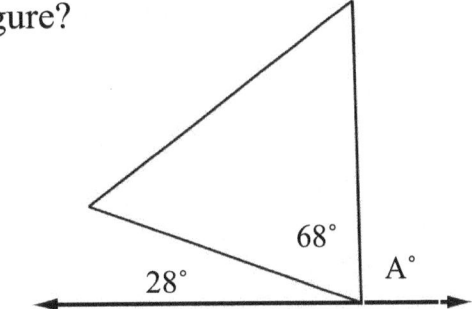

3) The temperature on Sunday at 12:00 PM was 102°F. Low temperature on the same day was 74°F cooler. Which temperature is closest to the low temperature on that day?

   A. 54°F

   B. 31°F

   C. 28°F

   D. 38°

MCAS Subject Test Mathematics Grade 4

4) Which statement about the number 342,597.61 is true?

   A. The digit 4 has a value of (4 × 1,000)

   B. The digit 2 has a value of (2 × 100)

   C. The digit 9 has a value of (9 × 10)

   D. The digit 3 has a value of (3 × 10,000)

5) What is the perimeter of this rectangle?

   A. 36 cm

   B. 104 cm

   C. 21 cm

   D. 42 cm

6) What is the eighth number in the following pattern?

   1,350; 1,500; 1,650; 1,800; ____; ____; ____; ____

   Write your answer in the box below.

7) Jamie has 6 quarters, 7 dime, and 35 pennies. How much money does Jamie have?

   A. 180 pennies

   B. 250 pennies

   C. 255 pennies

   D. 235 pennies

MCAS Subject Test Mathematics Grade 4

8) Order the fractions from least to greatest. $\frac{1}{2}, \frac{3}{4}, \frac{5}{8}, \frac{1}{16}$

   A. $\frac{1}{2}, \frac{1}{16}, \frac{3}{4}, \frac{5}{8}$

   B. $\frac{1}{16}, \frac{1}{2}, \frac{5}{8}, \frac{3}{4}$

   C. $\frac{1}{16}, \frac{1}{2}, \frac{5}{8}, \frac{3}{4}$

   D. $\frac{1}{2}, \frac{5}{8}, \frac{1}{16}, \frac{3}{4}$

9) Rounded to the nearest 10,000, the population of Louisiana was 74,920,000 in 2010. Which number could be the actual population of Louisiana in 2010?

   A. 74,816,420

   B. 74,929,650

   C. 74,923,335

   D. 74,849,920

10) What is the perimeter of the triangle?

   A. 95 inches

   B. 85 inches

   C. 90 inches

   D. 88 inches

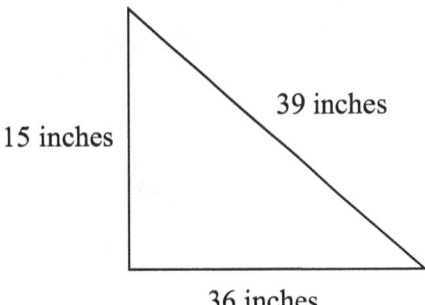

www.MathNotion.Com

MCAS Subject Test Mathematics Grade 4

Session 2

❖ **Calculators are NOT permitted for this practice test.**

❖ **Time for Session 2: 85 Minutes**

MCAS Subject Test Mathematics Grade 4

11) The figure below shows a diagram of a living room.

The perimeter of the living room is 58 feet (ft). What is the width(w) of the living room?

A. 22 ft

B. 21 ft

C. 24 ft

D. 26 ft

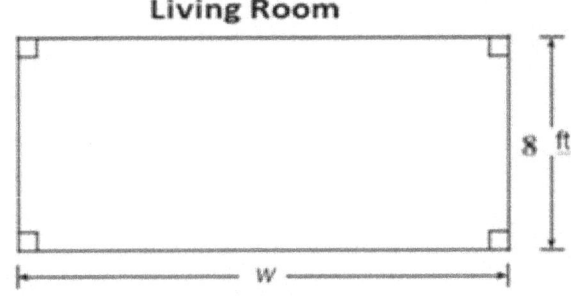

12) A building is 78 feet high. What is the height of the building in yards?

Write your answer in the box below.

13) Which triangle has one obtuse angle?

A.

C.

B.
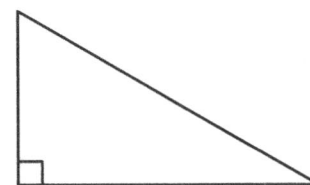

D. A and C is true

MCAS Subject Test Mathematics Grade 4

14) What is the volume of the cube?

   A. $600 \, m^3$

   B. $300 \, m^3$

   C. $1,000 \, m^3$

   D. $3,000 \, m^3$

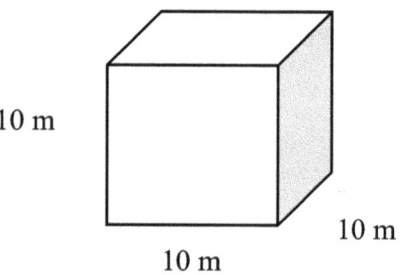

15) What is the perimeter of this shape?

   A. 28

   B. 17

   C. 20

   D. 25

16) There are 365 days in a year, and 24 hours in a day. How many hours are in one-fourth years?

   A. 3,240

   B. 2,910

   C. 4,280

   D. 2,190

MCAS Subject Test Mathematics Grade 4

17) A football teams play 74 games each year. How many games will the team play in 5 years?

   A. 1,320

   B. 350

   C. 370

   D. 1,420

18) Which shape shows a line of symmetry?

   A.

   C.

   B.

   D.

19) On Sunday Leon was a referee at 4 soccer games. He arrived at the soccer field 20 minutes before the first game. Each game lasted for 20 minutes. There were 10 minutes between each game. Leon left 15 minutes after the last game. How long, in minutes, was Leon at the soccer field?

   Write your answer in the box below.

20) Joe put 13 red cards and 11 black cards in each bag. What is the total number of cards Joe put in 9 bags?

Write your answer in the box below.

☐

"This is the end of Practice Test 2."

MCAS Subject Test Mathematics Grade 4

Chapter 13 : Answers and Explanations

Massachusetts Comprehensive Assessment System Practice Tests

Answer Key

❈ Now, it's time to review your results to see where you went wrong and what areas you need to improve!

MCAS - Mathematics

Practice Test - 1				Practice Test - 2			
1	B	11	B	1	A	11	B
2	4	12	C	2	C	12	26
3	C	13	D	3	C	13	D
4	C	14	9	4	C	14	C
5	C	15	B	5	D	15	D
6	9	16	B	6	2,400	16	D
7	B	17	C	7	C	17	C
8	B	18	C	8	C	18	A
9	B	19	A	9	C	19	145
10	A	20	7,000	10	C	20	216

MCAS Subject Test Mathematics Grade 4

Practice Test 1
MCAS- Mathematics
Answers and Explanations

1) Answer: B.

The clock shows 11:15 in the morning. 3 hours ago, it was 8:15 AM. 20 minutes before that was 7:55 AM.

2) Answer: 4.

7 days = 1 week. 31 days = (31 ÷ 7) = 4.28 ⟹ 4 weeks

3) Answer: C.

The football team should buy 12 uniforms and each uniform cost $16. Therefore, they should pay (12× $ 16) $192.

Choice C is correct answer: (16 × 10) + (16 × 2) = 160 + 32 = $192

4) Answer: C.

The number of votes for Football was 11, for basketball was 2, for soccer was 7, and for volleyball was 6. Only table C shows all these numbers correctly.

5) Answer: C.

32 + Z = 85. Therefore, Z = 85 – 32 = 53. Let's review the options provided.

A. 85 – 32 = Z   Yes! This is true. 85 – 32 = 53

B. 85 – Z = 32   Yes! This is true. 85 – 53 = 32

C. Z – 32 = 85   No! This is not true. 53 – 32= 21

D. Z + 32 = 85   Yes! This is true. 53 + 32 = 85

Option C is the only option that is NOT true.

6) Answer: 9.

Tam wants to divide his 720 cards into boxes of 80 cards. Therefore, he needs 720 ÷ 80 = 9 boxes.

7) Answer: B.

Choice B shows the numbers in order from least to greatest.

12,850< 21,080< 23,031< 28,017

MCAS Subject Test Mathematics Grade 4

8) Answer: B.

8.0 − 2.55 = 5.45

9) Answer: B.

Add adult tickets and children's ticket that have been sold.

$\frac{18}{100} + \frac{7}{10} = \frac{18}{100} + \frac{70}{100} = \frac{18+70}{100} = \frac{88}{100} = \frac{22}{25}$

10) Answer: A.

This angle is less than 90°. Only choice A shows angle less than 90°.

11) Answer: B.

Only on option B, all shapes have at least one set of parallel side and one set of perpendicular side.

12) Answer: C.

For the height of 1 centimeter, we have 4 pennies, therefore, for the height of 9 centimeters, we have 4 × 9 pennies.

13) Answer: D.

The first model from left is divided into 6 equal parts. 4 out of 6 parts are shaded. The fraction for this model is $\frac{4}{6} = \frac{2}{3}$. The second model is divided into 8 equal parts. 4 out of 8 parts are shaded.

Therefore, the fraction of the shaded parts for this model is $\frac{4}{8} = \frac{1}{2}$. These two models represent different fractions.

14) Answer: 9.

$45 \div A = 5$, therefore, $A = 45 \div 5$ $A = 9$

15) Answer: B.

$\frac{77}{100}$ is equal to 0.77.

16) Answer: B.

56 ÷ 7 = 8.

17) Answer: C.

Only option C fits Emma's description.

MCAS Subject Test Mathematics Grade 4

18) Answer: C.

24 × 14 = 336.

19) Answer: A.

If one box has 80 flashcards, therefore, 20 boxes have the capacity of (20 × 80) 1,600 flashcards.

20) Answer: 7,000.

When rounding to the nearest thousand, you will need to look at the last three digits. If the last three digits is less than 499 rounds down the number ending with three zeros. On the other hand, if the last three digits are 500 or more, round up the next ending with three zeros. Since, in the number 6,708 the number 708 is bigger than 499, then round the number to 7,000.

MCAS Subject Test Mathematics Grade 4

Practice Test 2
MCAS - Mathematics
Answers and Explanations

1) Answer: A.

1 issue of the magazine = $6

$168 ÷ $6 = 28 issues of the magazine

2) Answer: C.

Three angles in a triangle add up to 180°.

A° + 68° + 28° = 180° ⇒ A° = 180° − 96° = 84°

3) Answer: C.

Low temperature is 74°F cooler than the temperature at 12:00 PM which is 102°. Low temperature is 28°F (102°F − 74°F) that is choice C.

4) Answer: C.

A. The digit 4 has a value of 4 × 10,000, not 4 × 1,000.

B. The digit 2 has a value of 2 × 1,000, not 2 × 100.

C. The digit 9 has a value of 9 × 10. This is true!

D. The digit 3 has a value of 3 × 100,000, not 3 × 10,000.

5) Answer: D.

use perimeter of rectangle formula. P = 2 (length + width)

P = 2 × (8 + 13) = 2 × 21 = 42 cm

6) Answer: 2,400.

The difference of each two successive numbers is 150.

Add four 150 to last number (1,800):

1,800 + 150 + 150 + 150 + 150 = 2,400

7) Answer: C.

6 quarters = 6 × 25 pennies = 150 pennies

7 dimes = 7 × 10 pennies = 70 pennies

In total Nicole has 255 pennies

MCAS Subject Test Mathematics Grade 4

8) Answer: C.

To compare fractions, we can write fractions with the same denominator. Then, we can compare the numerators of each fraction and put them in correct order from least to greatest or greatest to least.

Common denominator of 2, 4, 8 and 16 is 16. Rewrite the fractions:

$\frac{1}{2} = \frac{8}{16}$ $\frac{3}{4} = \frac{12}{16}$ $\frac{5}{8} = \frac{10}{16}$ $\frac{1}{16} = \frac{1}{16}$

9) Answer: C.

To round numbers to the nearest ten thousand, make the numbers whose last four digits are 0001 through 4999 into the next lower number that ends in 0000. For example, 54,424 rounded to the nearest ten thousand would be 50,000. Choice C is correct because last four digits of 74,923,335 is less than 4,999.

10) Answer: C.

To find the perimeter of a triangle, add all three sides of the triangle $\Rightarrow$ P = 15 + 36 + 39 = 90 inches.

11) Answer: B.

Use perimeter of rectangle formula. P = 2 × (Length + Width)

58 = 2 × (8 + W) $\Rightarrow$ 58 = 16 + 2 × W $\Rightarrow$ 2 × W = 58 – 16 = 42 $\Rightarrow$ W = 21 feet

12) Answer: 26.

3 feet = 1 yard;   78 feet = 26 yards

13) Answer: D.

An obtuse triangle is one with one obtuse angle (greater than 90°) and two acute angles. Since a triangle's angles must add up to 180°, no triangle can have more than one obtuse angle.

In this term shape A and C have one obtuse angle.

14) Answer: C.

To find the volume of cube, multiply one side of the cube by itself 3 times:

Volume of a cube = (side) × (side) × (side) = 10 × 10 × 0 = 1,000 m^3

MCAS Subject Test Mathematics Grade 4

15) Answer: D.

To find the perimeter of the shape, add all four sides.

P = 4 + 10 + 4 + 7 = 25

16) Answer: D.

1 year = 365 days, 1 day = 24 hours

$\frac{1}{4}$ years = $\frac{1}{4}$ × 365 × 24 = 2,190 hours

17) Answer: C.

1 year = 74 games

5 years = 5 × 74 = 370 games

18) Answer: A.

You can find if a shape has a Line of Symmetry by folding it. When the folded part sits perfectly on top (all edges matching), then the fold line is a Line of Symmetry. Shape A shows a line of symmetry.

19) Answer: 145.

Each game 20 minutes, therefore 4 games took 4 × 20 = 80 minutes.

10 minutes between each game. There are 30 minutes in total between 4 games. (between game 1 and 2, 10 minutes and between game 2 and 3, 10 minutes and between game 3 and 4, 10 minutes)

Leon arrives 20 minutes before first game and left 15 minutes after the last game.

In total, he was 80 + 30 + 20 + 15 = 145 minutes at the soccer field.

20) Answer: 216.

1 bag = 13 red cards + 11 black cards (13 + 11 = 24 cards)

9 bags = 9 × 24 = 216 cards

"End"

www.ingramcontent.com/pod-product-compliance
Lightning Source LLC
Chambersburg PA
CBHW080442110426

42743CB00016B/3251